Make Your Mark in Science

D0166515

Make Your Mark in Science

Creativity, Presenting, Publishing and Patents

A Guide for Young Scientists

Claus Ascheron and Angela Kickuth

A JOHN WILEY & SONS, INC., PUBLICATION

Copyright © 2005 by John Wiley & Sons, Inc. All rights reserved.

Published by John Wiley & Sons, Inc., Hoboken, New Jersey.
Published simultaneously in Canada.

Library of Congress Cataloging-in-Publication is available.

ISBN 0-471-65733-6

Printed in the United States of America

10 9 8 7 6 5 4 3 2 1

Contents

Preface

School and university science courses generally prepare students well for the challenges of understanding science, carrying out experiments, and interpreting results. Surprisingly, however, the topics of creativity in science and science communication are only rarely a part of the curriculum of science courses at colleges and universities. Most young scientists are left simply thrown in the deep end and expected to learn by trial and error. But, as is amply evident from the working habits, the presentations, and the publications of many senior scientists, that learning process often comes to a halt at a fairly rudimentary stage.

Furthermore, having left the fold of the teaching environment, young researchers often struggle for several years to find their feet and establish themselves within the larger scientific community. The purpose of this book is to support young scientists during this early phase of their career and, by giving the relevant advice and information in a concise form, to speed up this learning process.

This book will not teach you any science. Instead, it addresses the often neglected questions of (1) how to work creatively once you have left the exams and course books behind you, (2) how to effectively communicate your scientific achievements in oral presentations and written publications and (3) whether and how to protect your accomplishments with a patent. If taken to heart, the advice given will also help you to adopt good working habits from which you will benefit throughout your career in science. In addition the book gives much background information about the machinery of scientific publishing and also about the modern possibilities of electronic publishing. A further chapter is devoted to patents – since this is an area in which scientists frequently fail to recognize and exploit good opportunities.

The main beneficiaries of this advice and information are likely to be graduate students and other young scientists; but anyone lacking confidence in these activities will find useful hints for improving their creativity and effectively mastering the task of communicating their ideas and results. Even senior scientists and professors with years of teaching experience might profit from reviewing their style of working or the quality of their oral presentations.

The style of this book is informal and easy to read, and intricate technical details are avoided. There are a number of other texts on scientific writing and presenting in which the reader can find all conceivable facts, figures and conventions. In contrast to such detail-laden tomes, our aim when writing this text was to make it concise, enjoyable to read and easy to digest. We hope we have succeeded in this aim.

For elucidating the topic of effective science communication from all relevant perspectives, the authors are able to draw on a wide spectrum of experience, including:

▷ scientific writing in numerous international scientific journals;

▷ lecturing and teaching experience in various countries and at various levels;

▷ assisting colleagues and students in preparing their papers and presentations;

▷ assessment of countless book manuscripts, journal articles and oral presentations during their work as editors at a leading scientific publisher.

One of the authors (CA) has been invited by many institutes and universities to lecture on the topics of this book. A recurring comment at the end of these lectures was: '*It would be great to have this information available in the form of a book too*'. So here it is. We hope you will enjoy reading it and that you will be able to use it to your advantage.

Acknowledgements. The authors would like to thank Bettina Posselt and Gerhard Mühlbauer for reading the complete manuscript and providing valuable comments and corrections. Mike Edmeades kindly contributed expert advice on the topic of patents. Our immense gratitude goes to Michael Ascheron for his tireless work in formatting the text, and also to Katharina Ascheron for serving as a 'model presenter'. Finally, we thank our respective spouses Regina Ascheron and Rolf Kickuth for their patience and support during the many months in which we were preoccupied with the job of writing, re-writing and re-re-writing this text.

Heidelberg, October 2004 *Claus Ascheron*
Angela Kickuth

1 Introduction

Becoming a well known scientist and achieving great breakthroughs is the secret dream of almost every enthusiastic young science student. Some – albeit a small minority – will go on to fulfil this dream. This is not to say that all scientists are motivated predominantly by the desire to attain personal fame and glory. Whilst this is a component of most young scientists' ambitions, there are other equally important elements: one is simple curiosity and the wish to understand nature; another is the desire to make a contribution to the great edifice of man's scientific knowledge of nature. Whatever your primary motivation, one thing is for sure: In order to become a successful scientist, you will need to possess the necessary qualities and skills.

1.1 What Does It Take to Be a Successful Scientist?

The personal qualities that are a prerequisite for being good at science can be readily identified. They include:

- a genuine curiosity and desire to explore and understand the world around us
- the ability to think and reason logically
- an aptitude for mathematics (especially in physical science)
- an adequate ability to memorise facts.

If you possess these qualities and have successfully completed a corresponding course of study, then you have the potential to make your mark in science. But this potential will remain latent, unless accompanied by another group of skills. These skills, which – in contrast to the above-listed qualities – can largely be learned are:

▷ creativity and a creative working style

▷ presentation skills

▷ writing and publishing skills.

It is to this second group of skills and the relevant background information that this book is dedicated. From its pages you will not learn to be a better scientist, but – by maximising your latent potential – you may well become a more successful scientist!

1.2 Creativity

A major challenge and barrier for many new research scientists is the need to work imaginatively and creatively: to have ideas for a new research project, or to find a new way of tackling an old but unsolved problem. During a typical science degree course, students acquire immense amounts of knowledge and learn to understand complex ideas and interrelations. But to a large extent they are spoon-fed from a curriculum and have neither the need nor the motivation to come up with truly new ideas.

Many people believe that creativity is a gift that you are either blessed with or not. Creative ideas – some would claim – fall from heaven into the minds of those who are properly receptive.* We do not agree! Our view is that everyone has the ability to work creatively, provided they adopt the right attitude and can work under suitable conditions to facilitate the creative process. In Chap. 2 we begin by discussing the nature of creativity, especially as revealed in science. Thereafter we suggest various ways in which you can enhance your creativity by choosing, among other things, optimal working conditions and life style. This chapter also includes forays

---

* Interestingly, the German word for idea/inspiration is Einfall, meaning literally '*a falling in*'. This supports the supposition that traditionally, ideas were thought to fall into the mind from an external source

into the related topics of mental flexibility (crystalline/fluid intelligence) and a look at factors that reveal and account for international variations in creativity.

1.3 Presenting

Having worked creatively to produce some interesting scientific results, the next challenge for a young researcher is to communicate these to a wider audience. If you keep your discoveries to yourself you will do nothing to promote your own standing as a scientist, nor will you contribute to furthering your field of research.

Your first formal opportunity to tell others about the results of your work will probably come in the form of an internal seminar or a short talk at a conference. Some people – but a minority – have a natural gift for presenting. They possess an easy manner, the ability to think on their feet, and can readily establish a rapport with their audience. Others are fortunate enough to have good mentors, whose style they can copy and whose constructive criticism helps them to succeed from the start. But most scientists have to learn the hard way, suffering the repeated criticism of colleagues, and ending their first talks with the feeling that the audience has completely lost interest. Such learning-by-doing may work to some extent, but it can be a long and painful process and has no guarantee of success.

Good presentations, in the form of lectures, are also enormously important in science teaching. All of us have suffered through lectures that were incomprehensible, inaudible and illegible and have then gone away, no wiser than before, to learn the same material from a textbook. In other words, not even all senior scientists and lecturers have mastered the techniques of good presenting. After many years of making the same mistakes, these habits are usually so ingrained that the perpetrator is no longer aware of them. Thus it is a worthwhile investment for young scientists to try to perfect their style of presenting from the outset. Chapter 3 of this book will help you to avoid the most common errors and guide you in preparing and delivering convincing presentations.

1.4 Publishing and Electronic Publishing

In practice, the most direct motivation for publishing and presenting your scientific findings is to document your contribution and to establish priority of discovery. It is a fact of life that scientists usually become known to, and accepted into, the scientific community through their scientific publications (*'publish or perish'* as the saying goes) and, to a lesser extent, as a result of their conference presentations. Your papers and your presentations – if they stand out from the crowd – are what get you noticed by your peers and seniors.

Writing and publishing good quality papers is thus another essential activity for young graduate or postgraduate scientists trying to establish themselves as fully-fledged researchers. Only a few manage to enter the scene with a bang, as did Josephson or Mössbauer, for example, each receiving the Nobel Prize at a relatively young age. For the majority of young scientists, the first years, if not decades, of their career involve a patient process of gradually advancing their status, often with disappointments along the way. At first, they usually have to share the recognition for their discoveries with senior colleagues. It can be a long time before they are able to produce a first-author or single-author paper. But this may not be a bad thing. These early years provide the best opportunity for a young scientist to learn the tricks of the trade with regards to publishing and communicating his or her results. However, the learning process is often a very haphazard one.

As mentioned in the preface, formal instruction on how to communicate your own scientific achievements is rarely included in university science courses. Nonetheless, writing good papers, like good presenting, is a skill that can be learnt. With the help of Chap. 5 of this book, you will rapidly come to know the main criteria defining well written papers and also learn to avoid the most common pitfalls. Your colleagues may be surprised to note a transformation in the quality of your publications. At some later point you should take the time to check your own style once more and to examine the style of colleagues in the light of the advice given here.

You will soon see what a difference it makes if you adhere to a few simple rules.

A good deal of Chap. 4 (Culture of Scientific Publishing), some of Chap. 5 (Scientific Writing) and all of Chap. 6 (Electronic Publishing) are included to give the reader some essential background knowledge about how modern publishing works: What kinds of publications are there? How do they differ? What new opportunities has electronic publishing brought? This additional information will be useful to all readers who have to deal with publishers and publications, but are not sure what options are available to them.

1.5 Patents

In our final chapter (Chap. 7) we discuss when and how to apply for a patent. One can find countless stories of excellent inventions whose originators have failed to gain patents (and have therefore lost out financially too) as a result of bad timing, premature disclosure of their ideas, or for other formal or legal reasons. Thus the aim of Chap. 7 is to help you recognize a patentable invention and decide what to do with it. Basic elements of patent law in America and Europe are explained, as is the application procedure.

2 Scientific Creativity

The desire and ability to create, whether in art, science or other fields of endeavour, is a uniquely human faculty. Indeed, together with language, it is the vital quality that has raised the human race to its unchallenged position at the pinnacle of biological evolution. In the words of Edward de Bono: *'There is no doubt that creativity is the most important human resource of all. Without creativity, there would be no progress, and we would be forever repeating the same patterns'.*

In this chapter we will begin by seeking a definition of creativity. We move on to consider the special features of creativity in science, thereby often drawing upon the opinions and experience of well-known creative geniuses. It will emerge that creativity in art and science have much in common. In the remaining sections of the chapter we will lay out some guidelines that should help you, the practising scientist, to optimize and exploit your own creative abilities.

2.1 What is Creativity?

Much has been said and written about the creative process, by scientists, sociologists and other academics. But nowhere have we encountered a clearer definition than that given by the British immunologist and Nobel Prize winner Sir Peter Medawar. In his 1985 essay* *'Creativity – Especially in Science'* he begins with the words:

> *'Creativity is the faculty of mind of spirit that empowers us to bring into existence, ostensibly out of nothing, something of beauty, order or significance'.*

---

* In Peter Medawar: The Threat and the Glory (Oxford University Press, 1990)

Ever since the time of Plato, many philosophers have held the romantic notion that creativity is of divine origin. In common with Medawar, and most modern thinkers, we do not subscribe to this view. Whatever the true roots of human creativity, it is a process that is highly vulnerable to error. This alone is strong evidence against a divine origin!

Creativity is clearly an inherently mental process. In essence it amounts to having a new and illuminating idea. In art the idea will be beautiful, moving or even shocking. In science it will be *'right'*, convincing, explanatory, or the solution to a longstanding prob lem. But the idea alone is not enough. It requires a medium of expression: For a painter or sculptor this medium is the physical material constituting the work of art; for a poet or musician the medium will be words or notes. For a scientist the vehicle consists of the words, diagrams and mathematics used to express relationships, to describe natural phenomena and explain observations. Whatever the tools of the trade, the creative process – whether in art or science – can only come to fruition when the creator also masters the use of these tools. Einstein, for example, actually had the idea for his General Theory of Relativity nearly ten years before publishing it. In the interim he had to learn and further develop the mathematics needed to formulate the idea in a manner that could be used to communicate it to other scientists.

2.2 Creativity in Science

In science, creativity is directly associated with gaining new insights into the workings of Nature, discovering as-yet unknown laws and relationships, explaining puzzling results and solving open problems. The type of thinking necessary to make such progress was described by Max Planck in his autobiography: '*A scientist must have a vivid intuitive imagination for new ideas not generated by deduction but by artistically creative imagination.*' The key phrase here is '*not generated by deduction*'. At this point it is useful to mention the concept of '*lateral thinking*', a term coined by de Bono. Lateral thinking involves seeking new ways to look at a problem, viewing it from new angles, and not merely proceeding by logical steps from what is already known or believed. It may even require seemingly illogical reasoning. As William Bragg Sr. put it: '*The important thing in science is not so much to obtain new facts as to discover new ways of thinking about them.*'

Scientific creativity, however, is not only the process of finding appropriate answers to puzzling questions; it is also the facility to ask the right questions at the outset.

2.2.1 On Inspiration

What is this mysterious ability that allows us to think laterally, to ask the right question, or to have the decisive idea? Such illuminating thoughts are often described as inspiration. For the more powerful and sudden insights of this kind, Arthur Koestler[*] – with due recognition to Archimedes – invented the term '*Eureka acts*'. As stated at the outset, most modern scholars no longer consider inspiration to be a divine process in which an idea is implanted in the mind by an other-worldly source. In fact, there is much evidence – concrete and anecdotal – to suggest that inspiration is largely the result of unconscious processes. The decisive idea emerges in the conscious mind, often at an unexpected moment, after a long period of gestation during which the brain has been ruminating, unconsciously, on the problem.

Inspiration of this kind has been attested to by artists and scientists alike. The composer Anton Bruckner, for example, woke up one morning with a sequence of notes in his head that he had just dreamed about. He jumped out of bed and wrote them down immediately (essential if you have dreamed something and want to remember it!). These notes became the core of the 3rd movement of his famous 4th symphony.

Henri Poincaré was very interested in the process of mathematical creativity and wrote[**] of one of his own breakthroughs:

[*] Arthur Koestler: The Act of Creation (1964, Hutchinson & Co.)
[**] Henri Poincaré: The Foundations of Science (The Science Press, 1913)

Fig. 2.1. Creativity as heavenly inspiration. A common misconception

'*Most striking, at first is this appearance of sudden illumination, a manifest sign of long, unconscious prior work. The role of this unconscious work in mathematical invention appears to me incontestible … There is another remark to be made about the conditions of this unconscious work: it is possible, and of a certainty it is only fruitful, if it is on the one hand preceded, and on the other followed by a period of conscious work. These sudden inspirations never happen except after some days of voluntary effort which has appeared absolutely fruitless and whence nothing good seems to have come, where the way taken seems totally astray. These efforts have not been as sterile as one thinks; they have set agoing the unconscious machine and without them it would not have moved and would have produced nothing.*'

As Poincaré recognizes, the brain can only engage in this subconscious activity if it has been set on track by prior conscious work. Beyond that, it must also be '*programmed*' with the relevant experience and have, where necessary, a broad base of knowledge on which to draw. Louis Pasteur expressed this nicely in the phrase '*Fortune favours the prepared mind*'. The subconscious process may also be enhanced by continuous close involvement with the endeavour concerned (although sometimes a holiday or change of scenery are needed to spark the breakthrough of an idea into the conscious mind – cf. Sect. 2.3.3)

In your own scientific work you have probably encountered this kind of inspiration, at least on a small scale: If you work very intensively on a particular scientific problem, it can possess your attention to such an extent that all your thoughts and everyday life comes to revolve around the problem. Even when you are superficially thinking about other matters, beneath the surface your mind is still ruminating on the problem. Your accumulated knowledge and learned pathways of logical thought allow the brain to continue the search for a solution, even in 'background mode'. It can then happen that you wake up one morning with the decisive idea in your head, i.e. the inspiration required to solve the problem. Expressed slightly differently, intensive long-term involvement in a particular field will also help you to develop a good intuition when tackling new problems. Intuition is not something magic: it results when the brain is able to draw upon a large reserve of accumulated and relevant knowledge. During the thinking process the conscious mind may not be aware of this knowledge, hence our feeling that a special faculty, that of 'intuition' or 'insight' is at work. Such intuition enables decisions to be made more quickly and more reliably. Creativity and success will follow if you enjoy your work and have the enthusiasm to apply yourself to it with sufficient intensity.

2.2.2 On Language

Without language humans would never have been able to exploit their creative potential, either individually or as a society. Alongside creativity, language is the most important human faculty. The advent of language some fifty thousand years ago triggered a huge acceleration in the pace of human evolution, one that ultimately led to civilization and culture as we know it, and also enabled us to engage in science and technology.

Whilst language is essential both as a crystallizer of thought and as the vehicle for communicating ones ideas, it can also be a trap. The fact that a certain word exists does not mean that there is necessarily an exact counterpart in Nature. Words were invented by humans, many of them at a time when their inventors did not understand (scientifically speaking) the full story behind the concept they were naming. Examples of words that have turned out to be not only tools but also snares or straight-jackets include *'time'*, *'space'*, *'mass'*, *'force'*, *'weight'*, *'ether'*, and *'corpuscle'* in the physical sciences. In psychology, one might mention the terms *'will'*, *'sensation'*, *'consciousness'*, or *'conditioning'*. Even mathematics has its own set of preconceptions, enshrined by words such as *'limit'*, *'continuity'*, or *'countablity'*. These words are not simple tags attached to a person or object – they hide within them a set of assumptions and the particular kind of logic that prevailed at the time of their invention. Truly creative scientists need to question even the words used to express their thoughts about Nature. Einstein would never have been able to transform Man's view of the Universe if he had accepted the two words *'space'* and *'time'* in the sense that they were generally understood by his contemporaries. His ability to probe deeper than others before him he once described to a friend in the following, characteristically modest way: *'When I asked myself how it happened that I in particular discovered the Relativity Theory, it seemed to lie in the following circumstance. The normal adult never bothers his head about space-time problems. Everything there is to be thought about, in his opinion, has already been done in early childhood. I, on the contrary, developed so slowly that I only began to wonder about space and time when I was already grown up. In consequence I probed deeper into the problem than an ordinary child would have done.'*

2.2.3 Creativity in the Daily Work of a Scientist

Creativity is relevant in science not only to major breakthroughs of the type associated with Einstein, Darwin, or Crick and Watson. Certainly it is these in which the *'Eureka act'* – the moment of sudden insight – is most dramatic. But what about the daily work of a modern-day scientist carrying out a typical research project? Here progress is more commonly made in a less dramatic way. However,

* C. Seelig: Albert Einstein (Europa Verlag, Zurich 1954)

small steps forward can in fact be equally creative and, ultimately, just as important for advancing our overall knowledge. Examples might be finding a new interpretation of unexplained results; devising an experiment to test an unproven hypothesis; or starting a new direction of research. It is not only geniuses, famous scientists and Nobel Prize winners who are capable of creative work. Every scientist can work creatively, although the degree of creativity realized will vary significantly among individuals. It is wise to recall the words of Thomas Alva Edison, certainly a creative and successful scientist. But he too recognized that '*Genius is one percent inspiration and ninety-nine percent perspiration*', i.e. that genius does not only mean having great new ideas; one must also have the perseverance to bring them to fruition.

One should not underestimate the importance of the small and routine tasks, often the ones that cause the perspiration. Whilst less spectacular than the big discoveries, in total they contribute the largest part to the overall creative process. Consider for example the preparation of a scientific paper or a lecture. Some scientists regard this type of work as burdensome and lacking in any creative component. But this is the wrong attitude. Planning a research project, or even a simple experiment, also involves creativity. So too does presenting the results of your research. If you think carefully about the best way of presenting a set of scientific facts, be these new results or well-established knowledge, and about the best way of convincing your audience of the correct interpretation, you yourself will often arrive at a new perspective. This may even lead to ideas for further investigations, can reveal gaps that still need to be filled and, at the very least, will help you to gain a better understanding yourself. The more carefully you think about a topic, the more relations and analogies you will find to other topics. In preparing a coherent presentation of scientific facts for an audience, you are putting together the pieces of a mosaic. The result seems more complex at first, but gradually you, and later your audience, will begin to see the larger picture that emerges from the combined pieces. This can lead to a new quality of insight and may thus indeed be a highly creative process.

2.2.4 Underexploited Creativity

Left unexploited, neither great mental leaps nor small creative steps will lead to productive achievement. Ideas need to be followed up using the tools of the trade, namely mathematics, experimental expertise, routine data analysis etc. Only in this way can your ideas be put to the test, verified (or falsified!), and placed on a firm foundation so that they become ripe for communication to others. A famous scientist who failed to exploit his creativity to the full was Leonardo da Vinci. Although he undertook a good deal of painstaking work and earned acclaim for his insights into human anatomy, he could actually have achieved a good deal more than he did. A genuine polymath, da Vinci was also a talented artist and engineer. It was his desire to be creative in new fields that often sidetracked him from completing projects he had begun. Much of his work on science and technology only became known after his death when it was reconstructed from unpublished notes he had kept during his studies (see Fig. 2.3).

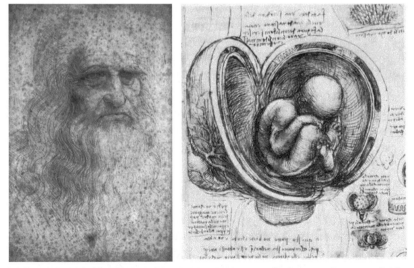

Fig. 2.2. Leonardo da Vinci

Fig. 2.3. A page from one of Leonardo da Vinci's notebooks

The moral of this story is that creative insights are themselves not enough. If you want your ideas to have the impact they deserve, then they must be adequately tested, the results interpreted and the full story written up and published, preferably in a high-impact journal (see Chap. 4). If you stop halfway towards this goal, you risk wasting your good idea, or at the least leaving someone else to receive the recognition by doing the necessary follow-up and publishing the correct interpretation. One can also cite more recent cases of researchers who were on their way to major breakthrough, but, failing to recognize the importance of their initial work, did not follow it up and allowed others to claim the glory. For example, Eiji Osawa from Toyohashi University actually calculated and predicted the existence of fullerenes some years before they were independently discovered experimentally. Had he sought – or even suggested – the experimental verification himself, he could well have shared in the 1996 Nobel Prize for Chemistry, awarded to Curl, Kroto and Smalley for their experimental discovery of Buckminster fullerenes or 'Buckyballs'. Exactly such a football-like

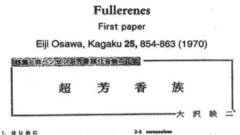

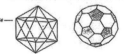

Fig. 2.4. The first paper on fullerenes – an example of creativity not fully exploited

structure was predicted by Osawa (see Fig. 2.4), but an additional factor hampering his chances of recognition was that he failed to published his theoretical predictions in an international English-language journal.

Thus creativity alone is not enough to guarantee productivity and success as a scientist. If you wish to achieve steady progress in your scientific work, you cannot simply rely on the rare moments of inspiration that come unannounced. For the most part you will need to engage in routine work, set up research projects, obtain results and write up the publications that will make your work known. Perseverance, patience and an element of intuition are the ingredients for success.

2.2.5 The Elusive Insight

We have all experienced the situation: Progress in a particular piece of work is hampered by a particularly stubborn problem. It may be an equation that needs to be solved, an experimental design problem, some unexpected experimental results that do not fit the theory, or a missing link in a logical chain of explanation. Whatever the problem, no amount of deliberation seems to bring the solution closer.

The best policy in such a situation, where your imaginative powers seem to be exhausted, is to turn your attention to other matters. You can still pursue other routine research work, set up projects, obtain further results, and take part in discussions. This is far preferable to brooding over the apparent brick wall blocking your progress.

In fact, the sought-after idea – especially if it involves a creative leap – will almost certainly not come to you while you are actively looking for it. A strong will to be creative is generally counterproductive when it comes to have '*big*' new ideas. Whilst a conscious desire to be creative can be useful for attaining goals that are already in sight, it cannot enable us to move in directions that have not yet been discovered. As Picasso once said '*I do not seek – I find*'.

2.3 Prerequisites for Creative Work

The potential for creative research work varies widely from one individual to another. Some people are born with a natural talent for devising novel ideas and automatically develop a style of working that nourishes creativity. Think for example of the composers Johann Sebastian Bach, Mozart, and Beethoven, the sculptor Rodin, the painters Rembradt and Rubens, or the scientists Marie Curie and Linus Pauling who each won two Nobel Prizes. But the majority, even of practising artists and scientists, have to make do with a more modest creative talent. But almost every researcher can learn to work with a satisfactory element of creativity if they adopt the right tactics.

It is also a fact that many great discoveries are made accidentally, a well-known example being Roentgen's discovery of X-rays. But had you been in the same position, would you have recognized the significance of Roentgen's chance observation, i.e. the mysterious blackening of his photographic plates? Even accidental discoveries can only be made if the scientist involved is open to the unexpected and able to pose the right questions. The information and suggestions in the following sections are designed to help you become mentally and organizationally prepared to exploit whatever creative opportunities fortune brings your way.

2.3.1 Mental Attitude and Personal Qualities

A very important prerequisite for creativity is the right mental attitude. Hand in hand with this go certain personality traits that are often observed among highly creative people. Qualities that seem to be particularly prevalent include:

▷ Diligence, or working for the sheer joy of it
▷ Stubbornness, or not listening to the chorus of '*you're crazy, it'll never work*'
▷ Ambition, or being filled with the need to prove oneself
▷ Eccentricity, or not caring about convention or conformism.

These, it should be stressed, are not necessary (nor sufficient!) qualities for creativity. Many very creative people lead normal and in other ways unremarkable family lives. But should you happen to possess one or more of the above qualities – don't worry, it may help you to reach that big breakthrough one day.

The Japanese Nobel Prize winner Leo Esaki identified the following key factors that he believes are conducive to creativity and will help you to reach the top as a scientist:

▷ Be open to the unconventional.

▷ Gain independence of the authorities, e.g. of your boss or professor.

▷ Avoid unnecessary information as it wastes time.

▷ Do not try to avoid confrontations.

▷ Stay curious.

▷ Be lucky.

One of the most important criteria is a natural curiosity about the world around us, i.e. the interest and will to discover something new for yourself. This has often been the source of a truly novel discovery. You should not be content simply to follow the paths that others have trodden before. But to go in new directions you need curiosity – to seek new insights, to ask questions that have not been

Fig. 2.5. Marie and Pierre Curie, Linus Pauling

asked before, and to think independently of the conventional wisdom. If a scientist only follows the ideas of his seniors or predecessors, he will at best be able to contribute to a gradual evolution of scientific knowledge. To make a major breakthrough or establish a new field of research something more radical is needed, a new and unconventional way of thinking.

Scientific geniuses such as Einstein could see things that no one else could see, were able to look at the Universe with fresh eyes. The way to do this – although hard to achieve – is to cast off preconceptions and empty your mind of all but the most basic information. Knowledge and experience are good for drawing analogies, for suggesting feasible solutions, etc., but – as recognized by Esaki too – knowledge overload is more likely to hamper creative thought.

2.3.2 Background Knowledge

Although a cluttered mind and too many fixed preconceptions may inhibit the process of free creative thought, it is clear that a good understanding of the foundations of your field will enable you to contribute most effectively to incremental progress. Many new ideas and inventions can actually be identified as a combination of existing ideas (which, however, need not detract from their importance). In order to recognize cross-links and analogies between different areas of science, it is valuable to have a wide background knowledge. This will enable you to transfer concepts and interpretations in a creative way to the topic of immediate interest. Typical examples of the transfer of concepts from one field to another are the application of methods of stochastic physics to financial markets, and the use of theories derived in fluid mechanics to model traffic flow.

Likewise, reading the relevant literature will enable you to keep a mental picture of the *state-of-the-art* in your field. This continually changing picture will help you recognize what new directions are possible and perhaps also some that are impossible, i.e. have already been explored and found to be dead ends. In this way the information gleaned from the literature should serve both as a fertilizer and a filter for your own deliberations.

The cumulative scientific knowledge and technical expertise of the human race as a whole is also a factor influencing the pace and timing of scientific breakthroughs. It is a puzzling but recurrent phenomenon in the history of science that the same significant discovery is often made more or less simultaneously by two or more scientists working independently. In 1922 Ogburn and Thomas published some 150 examples of discoveries and inventions made independently by several people. A well known example is the independent invention of calculus by Leibniz and Newton. These simultaneous discoveries are usually a reflection of the fact that scientific knowledge and technical feasibility have reached the prerequisite stage for the new idea or invention to emerge. It then simply requires the appropriate creative individual to come along and synthesize the parts into the new whole. Whilst suggesting a kind of inevitability about scientific progress, this observation should not be construed as reducing the role of the inventor from that of hero to midwife, assisting the inevitable birth. There are counterexamples too: Einstein discovered the principle of relativity unaided by any observation that had not been available for at least 50 years. The idea was ripe, but nobody – until Einstein – had been able to harvest it.

2.3.3 Stress Versus Relaxation

Pressure and stress at work; too much routine work; tension in social relationships: These are all factors that are counterproductive to scientific, or indeed any kind of creativity. A conscious effort to minimize all these factors may prove to be worthwhile. While minimizing such negative influences is helpful, one should not forget to promote positive factors. Since creativity is enhanced by a relaxed atmosphere and suppressed by a stressful environment, it is important to have some fun at work. Social contacts at work, in-

Fig. 2.6. Pressure and overwork will stifle creativity

cluding shared leisure activities such as sport, can also help to promote scientific output. A moderate level of stress is also desirable. However, it should be the positive kind of stress that results from working hard and playing hard. This type of stress, which is generally self-imposed, can indeed enhance your productivity. By setting yourself the goal of accomplishing a certain task within a certain period, you will be motivated to concentrate on it and to avoid distractions. A slight tension of this type is a prerequisite for activity.

If, however, the pressure of work is too high and, in particular the pressure is imposed on you externally and is beyond your control, then stress turns into distress. Putting in a 14- or 16-hour working day to keep up with the demands placed on you is not the answer. Under such circumstances scientists can suffer from burnout and their creativity will plummet. Thus a balance must be found between intensive working time and creative relaxational time. Sometimes the two can be combined – for example when you spend an evening out with your colleagues and the conversation includes both '*shop*' and other topics.

Sleep, too, should not be neglected. If you suffer from permanent overtiredness, creative moments will pass you by. But not only this: it has been shown recently that sleep plays an active role in promoting the creative process. During sleep and dreaming, information absorbed during the day is sorted, newly organized and correlated with information already stored in the long-term memory. After a period of sleep, test persons demonstrate a better problem-solving ability than those who have been awake in the interim.

Sufficient leisure time and sleep will help you to maintain an optimal state of mind – a relaxed mood in which you can concentrate well on the task at hand but also welcome new and unconventional ideas.

There is much truth in the old adage '*A change is as good as a rest*'. Creativity is certainly stifled by stress, but also by routine, and especially by routine thought patterns. To counteract the effect of getting stuck in a mental or physical groove, it is valuable to take regular vacations and engage in a variety of activities that are not related to your work. Often particularly conducive to creativity is a change of scenery.

For example, legend has it that Newton's idea for the law of gravity came to him as he lay in an apple orchard in England. An apple falling on his head is said to have given him the decisive inspiration. Better documented cases include Schrödinger, whose wave formulation of quantum mechanics was devised during a skiing holiday in Switerland, or Heisenberg, who had the idea for the uncertainty principle as he stood on the edge of a cliff on the island of Helgoland. Bohr's model of the atom apparently came to him whilst he was at a horse race and Galileo conceived the idea of using a pendulum as a clock whilst sitting in church.

2.4 Personal Working Conditions

We can't spend all our time on holiday, lying in orchards, or standing on cliffs looking out to sea. Indeed, this would have little chance of success if not accompanied by the necessary hard work and mental preparation. But there are plenty of other factors that you can quite readily influence, which, when optimized, can help you to increase the creative output of your daily work.

Not every factor mentioned below will be relevant to you, since each of us is sensitive to different aspects of our environment. But you will probably be able to identify a number of items that do indeed influence your peace of mind and interfere with your ability to work creatively.

Although much of your scientific work may be fairly routine, there is usually sufficient flexibility for you to arrange a working environment that reflects your personal preferences. Firstly, it should be noted that the number of hours spent working is irrelevant – it is the quality of the scientific work that counts. Thus, whenever you can, choose to work in a place where you can concentrate and think clearly. This may be in the office, at home, in the library or, on occasion, even on the aforementioned cliff-top.

It was the Americans who first came up with the idea of sabbaticals. Professors and other academics qualifying for a sabbatical are given paid leave from their university posts in order to concentrate on independent creative scientific work. This system has proved remarkably successful in terms of enhancing creative output.

2.4.1 Daily Routine

At what time of day do you concentrate best? Although this varies from one person to another, for those with a normal rhythm of waking and sleeping, the 'concentration curve' will generally be of the form shown in Fig. 2.8. It is apparent from this that the majority of us can concentrate best in the morning and again in the late afternoon. Thus the early afternoon and late evening – when most of us are less alert – are the times for dealing with routine work.

Regardless of your favoured time of day, it is not a good idea to fill every hour of the day with routine tasks and duties, leaving no time for unplanned creative activities. It is good to leave at least 10% of your time available for exploiting unscheduled and unexpected opportunities. Expect the unexpected. We all know people who always claim to be terribly busy, never allowing themselves to

Fig. 2.7. Library reading room – a place for creative thinking

be side-tracked into an '*irrelevant*' discussion. In our experience these are not the most efficient and best organized colleagues; and certainly not the most creative.

Organizing your daily work effectively, whilst leaving time for creative pauses, also demands that you learn to set priorities: Not every task is of equal importance or urgency. A more sophisticated style of working in which you tackle different tasks in parallel (delegating where appropriate!), is much more efficient than a linear style in which tasks are placed in a queue and tackled in turn. Recommendation: Review your list of outstanding tasks at least once a week, if not every day, and take a few minutes to assign priorities. Always be prepared to modify your priorities when circumstances change. In general, the least time-consuming tasks should be tackled first – this helps to give you a feeling of achievement and to avoid a long and discouraging list of things still to be done. Try to regularly set aside some time for more creative work, if necessary by delegating more work and responsibility to colleagues and juniors. Be aware that you cannot increase the amount you achieve without limit simply by extending your working hours. After a point this becomes counterproductive. No-one has ever said on their death-bed '*I wish I had spent more time at the office*'!

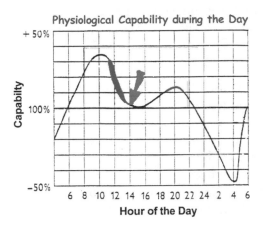

Fig. 2.8. Curve showing how our average ability to concentrate varies according to the hour of day

2.4.2 Environment and Lifestyle

The '*microcosm*' also has a significant influence for many scientists. Here individuals vary greatly and so we will simply list some of the factors that may be pertinent.

▷ Where do you have your best ideas? Some scientists can concentrate better in an undisturbed home environment, whereas others need the familiar environment of their workplace in order to '*switch on*' to their research.

▷ Some scientists have all or most of their best ideas during their working hours. Others find that such ideas come to them when jogging, walking in the country, relaxing in the bath (remember Archimedes!) or when lying in bed half-awake first thing in the morning. On occasion I (CA) have spent a whole day writing up in detail the stream of ideas for projects that came to me within the first 20 minutes of waking. Many readers will recognize this phenomenon.

▷ Being healthy and in good shape physically undoubtedly affects ones mental performance. Jogging or visiting a health club are thus recommended leisure activities. This has been recognized by many companies who provide their staff with weights rooms and other sports facilities. As conventional wisdom confirms: *'mens sana in corpore sano'*, a sound mind (dwells) in a healthy body.

▷ Gentle physical exercise, and for some even endurance sport, can promote the flow of ideas. A long-distance run, swim or cycle ride can lead to a good harvest of ideas. The regular movement and breathing fully occupy the body, a circumstance that seems to allow the mind to wander off in interesting directions. Goethe is said to have been at his most creative when pacing up and down his room. In the process he dictated his manuscripts to his secretary Eckermann. Nowadays moments of inspiration, even fleeting ones, can best be captured by keeping a Dictaphone or at least a notebook with you.

▷ Travelling to work is regarded by some as wasted time. Edison, for example, avoided his short journey home by sleeping at his desk. In contrast, others have their best ideas during the walk or cycle-ride to the institute

▷ The more disorganized of us might claim '*A tidy desk is the sign of an empty mind*'. But to make the most of the time available it is really better to have the papers you need at your fingertips –

Fig. 2.9. As recognized by Einstein (seen here cycling in California), physical exercise can stimulate creativity

or at least to know where they are – rather than having to search for 10 minutes.

▷ The illumination of the office and of the desk can promote or hinder effective working. The colour of the light can be important, and neon light, in particular, tends to have a negative influence. A spotlight for reading and writing makes concentration on close work far easier over longer periods. Even the quality of the daylight can be important. Whilst intense sunlight on the desk is usually counterproductive, many find that the sunset and twilight have an inspiring effect.

▷ Even the temperature of the room can make a difference. For concentrated working in a seated position, an ideal temperature is 21–23° C.

▷ Fresh air with plenty of oxygen is preferable to stale air with a high CO_2 concentration, which induces tiredness.

▷ Many scientists prefer to work undisturbed and have an office to themselves. But in Japan, for example, professors are generally content to share a room with a secretary and nonetheless manage to concentrate on their scientific work. In American companies and institutes open-plan offices are common. Here each staff member has merely a cubicle to call their own. In a 1.5 by 1.5 m space equipped with desk, chair, computer and telephone, they are separated by a chest-high panel from the neighbouring cubicle. When accustomed to this environment most people can manage to concentrate without difficulty.

▷ Opinions and tastes differ widely concerning the effects of background sounds. Whilst one scientist may need complete quiet in order to concentrate, others get used to and come to need a background of traffic sounds, music or voices. Some companies have found that brainstorming sessions can be made more successful when the natural sounds of wind, waves or birds are present in the background. Other people claim to be able to concentrate better when music is played, especially light classical music.

▷ Pleasing sensory input can promote concentration and creativity. Whilst this is often visual, e.g. through having a nice view from the window or a favourite painting on the wall, music can also be beneficial (see above). Even the sense of smell can play a role. The German writer Friedrich Schiller liked to keep a rotting apple in the drawer of his desk while working. Well, each to his own! In fact, it is possible he was actually inhaling ethylene vapour produced by the apple, which could explain at least his perception that the apple made him more creative.

▷ This brings us naturally to the influence of coffee, tea, alcohol and other stimulants. The novelist Honoré de Balzac is said to have drunk up to 30 cups of strong coffee a night when working on his manuscripts. He resorted to the caffeine to stave off his desire to sleep. When he died at the age of 51, his doctor ascribed his death to a heart complaint, aggravated by the abuse of caffeine. But the coffee almost certainly enhanced Balzac's productivity: he completed 85 novels in just 20 years. Whilst it is known that several other famous literary figures, for example the poet Samuel Taylor Coleridge who drank freely and also became an opium addict, have allowed drugs to influence their creativity, documented cases in science are not known to us. Indeed, since science depends on rational thought and logic, it is unlikely that scientific creativity can be much enhanced through drugs.

Fig. 2.10. A novelist for whom caffeine was an essential aid: Balzac drawn by David (1843)

A cup of coffee to remain wakeful, or a glass of wine to relax perhaps, but beyond that?

▷ A good relationship with your family, friends and partner helps most people to remain mentally well balanced and productive. However, it is not uncommon for brilliant scientists to be 'loners'. So don't be ashamed or worried if you are not the life and soul of your institute's Christmas party. Whereas Richard Feynman was a very gregarious person who espacially enjoyed relaxing with pretty women, Henry Cavendish avoided close relationships with the opposite sex, apparently living only for his work.

▷ Although happiness and health will promote creativity for most of us, there are exceptions. Even severe physical and mental handicaps need not lead to a loss of creativity; sometimes just the opposite is found. Stephen Hawking would probably not have become the famous scientist that he is, had he not been confined to a wheelchair since his youth. Likewise the mathematician John Nash and the painter Vincent van Gogh both displayed formidable creativity despite suffering from severe mental disturbances. We do not mean to suggest that a handicap itself enhances creativity; but it should in no way deter those who are afflicted.

Fig. 2.11. Many of history's most creative geniuses have had close encounters with madness. The mathematician John Nash (born 1928 in Bluefield, West Virginia) overcame his schizophrenia to receive the 1994 Nobel Prize for Economics

This list is already long, but one can conceive of many more variables that can be adjusted to suit majority and individual preferences. To make the most of your potential you should seek the maximum overlap between your preferred working conditions and the opportunities available. You might even try asking your seniors to allow you to work at home or in an orchard every now and then. After all, it is in their interest too to optimise your efficiency and output.

2.5 Group Creativity: Cooperation and Competition

Conditions that can enhance creativity include both cooperation and competition with your fellow scientists. Competition can inspire you to work on a problem more intensely in order to solve it first. But, equally, an open and regular exchange of ideas will promote rapid progress in a developing field.

Probably the most important single source of new ideas are the personal exchanges between members of a research group, meeting in a relaxed atmosphere for coffee and spontaneous conversation. When all are enthusiastic about their research, it is natural that the talk should revolve around the common and individual challenges being addressed by the scientists, professors and students alike. This supposition was confirmed in a study commissioned by the top management of the company IBM. They were interested in finding out why scientists from the IBM research center in Rüschlikon, Switzerland had won several Nobel Prizes, whereas those at the similarly equipped and equally elite American IBM institutes had received correspondingly little recognition. The result pointed to the fact that the researchers at the Swiss center worked in a more relaxing atmosphere, meeting regularly in the cafeteria to chat with colleagues from other groups. This circumstance, it was supposed, helps to stimulate interdisciplinary exchange and the development of novel solutions to scientific challenges. Taking a break from the laboratory to chat to colleagues in the cafeteria can thus be even more important than single-minded concentration on the job at hand. At most research institutes it is indeed encouraged.

Meetings and collaborations with scientists from other institutes can also play an essential role in furthering your work. By exchanging ideas and combining different approaches to solving the same problem, new insights can be gained that neither party alone would have attained. Similarly, it is not unusual for the inspiration for a new research project to arise from presentations and discussions at conferences. Even though the information may only be partially absorbed during the conference, the exposure to ideas and results that are new to you may fuel your subconscious thinking and help you to have the decisive brainwave.

A famous case in which strong collaboration went hand-in-hand with intense competition was the development of quantum mechanics in the early years of the 20th century. At the time, only a few scientists were working on this problem, in particular Schrödinger and the groups in Copenhagen (under Niels Bohr), Göttingen (Born, Pauli, Heisenberg), Munich (Sommerfeld) and Cambridge (Dirac). As the result of regular written exchanges, many visits and, of course, some inspired individual ideas, together – over the space of 10 years – these scientists had jointly devised the entire quantum theory in much the form that is used today.

It fact, throughout most of the last century, many major discoveries and most small steps forward were made by groups of re-

Fig. 2.12. Quantum pioneers: Bohr, Schrödinger, and Heisenberg

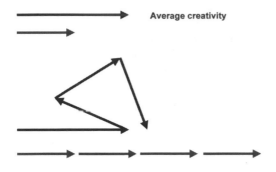

Fig. 2.13. Overall result of group creativity for two groups of scientists with different average creativity, but also different degrees of cooperation

searchers rather than individuals. This is partly the result of the technical complexity of modern research projects, especially experimental ones, but it is also an indication that we have moved on from the era of the great geniuses working in isolation to produce their breathtaking theories and explanations. Major progress today is favored by an active exchange of ideas and the combination of different ways of thinking about the same problem.

Let us consider two fictitious research groups. The first, whose members are individually not highly creative or headstrong, and the second, composed of more brilliant scientists each keen to follow up their own ideas, but less interested in the output of the group as a whole. As shown schematically in Fig. 2.13, the former group, in which each member contributes a small amount to the group objectives may well achieve more that the group of brilliant individuals with diverging interests. Note, however, that a highly creative and strong-willed group leader is desirable – not only to define the aims of the reasearch group, but also to motivate the team members.

A frequently fertile source of group creativity is provided by brain-storming meetings. It is advisable to first define some basic directions for the discussion and to assure that all participants are informed about these. Thereafter the discusssion should be no-holes-barred and should range from the practical via the unconventional to the fantastic. Each person should be encouraged to pick up and extend the ideas of the others. It is often an advantage if the discussion is guided, gently, by a clear-headed chairperson who is also able to keep a note of any good ideas that emerge.

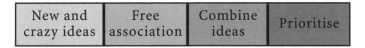

Fig. 2.14. Principles of Brainstorming

Finally, pushing the notion of competition even further, sometimes it can even be useful to have an intelligent '*enemy*'. If one of your colleagues is constantly trying to discredit your work, this will motivate you to think more carefully about what you do and say. It will help you to avoid weak points in your arguments and to develop and present a complete picture of your findings. Being forced to think more carefully may help you to have better ideas.

2.6 Intelligence and Creativity

Having considered the ways in which individual creativity can be influenced by working conditions and lifestyle, let us now take a look at the role of intelligence. This, as we will see, is a complex factor, which – to some extent – can also be influenced, both in individual and in team endeavours.

There is certainly a correlation between intelligence and scientific creativity. Most highly creative scientists are also highly intelligent in the conventional sense (high IQ). The reverse conclusion, however, is not valid. Many very intelligent scientists are not especially creative. They may be good problem solvers, e.g. they will solve that tricky equation if you write it down for them, but they may not be good at recognizing and formulating new problems to tackle.

In fact, the kind of intelligence associated with creativity cannot simply be characterized by the IQ figure measured in standard tests. Here we want to distinguish two main components of intelligence that contribute to scientific creativity: these we will call '*fluid*' and '*crystalline*' intelligence.

2.6.1 Fluid Intelligence

'*Fluid intelligence*' is the term we use to describe mental flexibility, i.e. the speed of understanding problems, of learning new content, and of interpreting situations. It also relates to the ability to draw conclusions from fuzzy information, to develop new ideas, to display correct intuition, to be alert and react quickly and appropriately in a new situation.

Faraday was a scientist with a very high degree of fluid intelligence and a natural gift for creative and intuitive thought. Without any formal training in mathematics, he discovered and correctly formulated a number of important physical laws.

A high level of fluid intelligence is found predominantly in young people, whose brains are known to be more '*plastic*' (this term, coined by neuroscientists, describes the greater ability of younger brains to form and reform connections between neurones).

As we get older, our brains tend to become less plastic. For a research scientist, one obvious way of maintaining mental flexibility is to have regular contact and discussions with students. In general, to promote and maintain mental flexibility, avoid allowing your

Fig. 2.15. Michael Faraday (1791–1867)

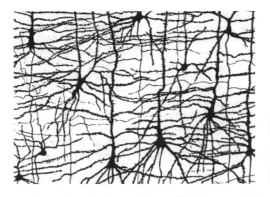

Fig. 2.16. Mental flexibility and the growth of new neurons (even in adults) is promoted by learning new skills

daily work to be routine only, remain open to new ideas, and be prepared to revise your views as a result of interactions with other scientists and students. Regular changes in task (called job enrichment in industry), forcing you to tackle new problems, will also help you to stay flexible. Learning something new, even though not related to your work (e.g. an instrument or a foreign language), is also an excellent way of promoting mental flexibility. Recent research has shown that, even in adults, new brain mass (possibly new neurons and/or new synapses) is formed when new tasks are learned.

In contrast, if, having learned your 'trade', you no longer attempt to learn anything new and if, furthermore, you interact only with your own generation, you are likely to develop rigid thought patterns and an attitude that is resistant to change and unwilling to entertain new ideas. This of course is the perfect recipe for stamping out creativity.

2.6.2 Crystalline Intelligence

If fluid intelligence were the only component, then we would all be going downhill from the age of about 25. Fortunately, however, for much of our lives what is lost through decreasing fluid intelligence is compensated for by growing 'crystalline' intelligence.

We use the term 'crystalline intelligence' for the component of intelligence commonly referred to as wisdom, i.e. accumulated knowledge and experience and the resulting '*feeling*' that one develops, not only for ones specialist topic, but also for life's challenges in general. Although it functions differently, relying on memories and fixed neural pathways, crystalline intelligence can often lead to the same results as fluid intelligence. By drawing on previous experience and finding analogies, it can help one to reach correct conclusions in new situations. Thus older scientists, with their greater store of knowledge, are often better at making strategic decisions than their younger colleagues. And although their output of new ideas may be somewhat lower, they are likely to be better at distinguishing between ideas with good potential and those that are non-starters.

Looking at Fig. 2.17 (the variation of fluid and crystalline intelligence with age), you may wonder how it is that many older professors still remain so active and productive. In fact, both in arts and science, older people often develop adaptive mechanisms that enable them to continue performing at a high level right into old age. Among these are selection, selective optimization, and compensation. The pianist Arthur Rubinstein enjoyed his greatest fame from the age of 47 onwards, only retiring in 1976 at the age of 89. Asked how he managed to remain such a virtuoso performer at his advanced age, he replied: '*I play fewer pieces* [selection], *practice them more frequently* [selective optimization], *and set more contrasts in tempi to make the playing seem faster* [compensation].'

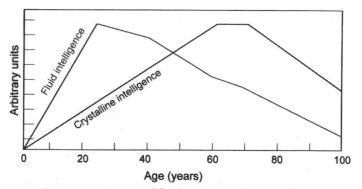

Fig. 2.17. Schematic variation of fluid and crystalline intelligence with age

2.6.3 Combining Fluid and Crystalline Intelligence

A scientist's overall creative potential is determined by the combination of fluid and crystalline intelligence. As seen in Fig. 2.18, the maximum of the *'intelligence/creativity'* curve lies somewhere in the forties. In maths and theoretical physics, scientists tend to peak somewhat earlier, since analytical skills and mental flexibility are paramount in these fields. In other disciplines, where accumulated knowledge and experience play a greater role, the creativity peak can be much later and the decline in output with age much slower.

Group creativity can greatly profit from the combination of fluid and crystalline intelligence provided by different team members. Such a mixed team can have a much greater creative output than the sum of the individual contributions. Particularly useful for realizing the intellectual potential of a team are *'brainstorming'* meetings (see also Sect. 2.5). By involving a number of independent minds – some young and very flexible, others older and highly experienced – such exchanges help to achieve one of the main prerequisites for creativity; namely the facility to view a problem from many angles. Although no-one knows at the outset what, if any, progress will emerge, excellent ideas have frequently been born in such, often chaotic exchanges.

2.7 Scientific Creativity and Productivity Worldwide

Creativity, whether in the arts or in science, is an entity that is very hard to quantify. There are no generally accepted means of evaluating creativity and – in modern art particularly – judgments about the creative value of a work are extremely subjective.

In science one can recognize that some individuals possess great creative talent, simply be considering their output of new and valuable ideas. But our aim in this section is to examine regional differences in scientific creativity and to seek explanations for these. The information in this section will not help you to develop your own creativity, but it might give you some critieria for deciding, for example, where to look for a post-doc position.

As a rather crude means of assessing creative output (perhaps better termed scientific productivity) we will look at some regional and national statistics relating to the following:

1) Number of papers published

2) Number of highly cited papers and highly cited researchers

3) Distribution of Nobel Prizes

4) Number of patents awarded.

We will consider these measures in turn in Sects. 2.7.2 – 2.7.5. But first let us make some general observations about the different working habits and expectations in different parts of the globe. These may help to shed light on the differences in scientific creativity in various regions. Our statements are necessarily based on gross generalizations, but they serve to show up some interesting trends. When it comes to the question of productivity – i.e. sheer volume of output – it will emerge that political and financial factors are the main source of the variation.

2.7.1 Conditions and Opportunities

Every scientist desires, and is expected, to produce good results and to publish them in a recognized international forum. Only by achieving such results and the corresponding highly cited publications can a scientist expect to progress to top-level positions in academia or industrial research. In the West, creativity is often observed to go hand-in-hand with a high degree of *'individuality'*, i.e. top scientists and professors often have rather unconventional approaches to their work, unusual working hours or environments, and many attach great importance to this freedom to choose how they work. Fortunately, most administrations at western research institutes accept or even promote this type of freedom for academics. They recognize that it is the scientific output that is important and not, e.g. presence in the office from 9.00 – 17.00. Universities, in particular, give their academics maximum space for creativity without the automatic pressure present in many companies. Sabbaticals too offer academics in the West an excellent opportunity to develop and exploit their creative potential.

However, a major source of pressure for academics in the West is that – these days, at least – many researchers, and perhaps the majority of young scientists, do not have permanent positions. They move from one temporary post to the next, and are only likely to get a new post if they produce outstanding results and numerous publications. This circumstance can persist for many years before a tenured position can be found. The pressure that it causes may well be just as extreme as that experienced in industry. This aspect of the western academic system may stifle some people's creativity. But for others, the competition that it generates actually contributes to a greater resolve to succeed. On balance it probably does not reduce the scientific output in relation to the financial input.

One can also try to trace the roots of international differences in creativity to the respective school systems. Although the average standards in US schools are lower than in many other western countries, N. America produces more than its share of top scientists. This can probably be attributed to the fact that, for the academic elite, the US provides excellent private schools, colleges and universities. It is not the average level that matters but the opportunities offered to the best students. One also finds such elite schools and universities in France and Great Britain; less so in Germany

Another country that plays a leading role in science and technology is Japan, although Japan – as the subsequent sections will show – does not produce as many outstanding individual scientists as, for example, the USA. Japanese children generally do very well in international comparisons of school performance and the universities enjoy the luxury of selecting only the most able students.

The probable reason for the lower number of '*star-scientists*' in Japan is the more rigid system of seniority and social interactions. Japanese citizens are brought up to view themselves as a small cog in a large machine; expression of individuality is not encouraged; and contribution to the overall enterprise is valued more highly than innovation.

This system, however, also has its advantages and certainly contributed to the dramatic success of Japanese industry in the 1980s, Companies were able to quickly implement new technologies and all employees worked together to give Japan a leading role in many branches of industry.

2.7.2 Number of Publications

As the first and perhaps crudest measure of scientific creativity/ productivity around the globe, let us consider the output of scientific publications from various regions. This is plotted, for the example of physics publications, in Fig. 2.18.

It is a remarkable fact that, independent of research area, about 70 % of all scientific publications originate from one of the three regions USA, EU, or Japan. Despite their size, countries such as China, Russia, or India make only a very small direct contribution. But this does not mean that Chinese, Russian and Indian scientists are not productive. The fact is that very many capable scientists from these and other developing nations choose to work abroad, often in America or western Europe, and their work contributes to the output ascribed to their country of residence. This '*brain drain*' has given cause for concern in various countries at various times. In the 1970s and 1980s the British feared that they were losing many of their most capable young academics to attractive positions in the USA. Today, the Chinese are trying to reverse the flow by offering professorships in China to Chinese scientists working in the US, and, most importantly, matching the salary that they receive in the US.

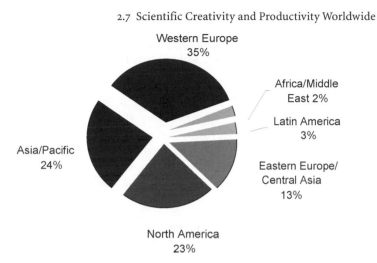

Fig. 2.18. Percentage of physics PAPERS in international journals from different regions of the world

These observations support the idea that the financial resources available to support research work and pay scientists' salaries are critical in determining the level of output worthy of publication in leading international journals. It is not enough just to have good ideas: frequently one also needs resources to realize the ideas. Furthermore, many scientists are only stimulated to have good ideas by virtue of being involved in a large and interesting research project. So it is not surprising that the elite US universties such as Princeton, Stanford, and Harvard, who offer the best salaries and research budgets, attract the most brilliant and motivated researchers, which in turn leads to a highly stimulating environment.

2.7.3 Highly Cited Publications

Another, perhaps more representative, measure of scientific productivity can be gained by considering the number of highly cited publications due to an individual, an institute or a nation. The number of highly cited, or '*high-impact*' papers clearly gives a measure not only of the volume of output, but also of its scientific

Table 2.1. Institutes producing the most high-impact papers in various fields of research (based on the statistics of the Institute for Scientific Information 1991-2001)

Area	Rank	Institution
Physics		
	1	Bell Labs
	2	Tokyo University
	3	IBM
	4	MIT
	5	CERN
Chemistry		
	1	UC Berkeley
	2	Kyoto University
	3	Tokyo University
	4	University of Texas
	5	Cambridge University
Engineering		
	1	MIT
	2	UC Berkeley
	3	NASA
	4	University of Illinois
	5	Stanford University
Materials Science		
	1	Tohoku University
	2	IBM
	3	UC Santa Barbara
	4	MIT
	5	University of Illinois

significance. Here one sees (Table 2.1) that the institutes producing the most high-impact papers are not confined to the US. European and Japanese institutes also feature prominently.

It should be noted that a highly cited paper does not necessarily report on highly creative work. On the contrary: very innovative work is rarely highly cited, for the simple reason that the ideas are new and few people are able to connect them quickly to their own work. It is often the more mundane mainstream papers, reporting a small advance in an active and popular field that receive the most citations. Consistent with this view is the fact that only three Nobel Laureates are among the 100 most cited physicists.

2.7.4 Nobel Prizes

If one were to consider only the international distribution of Nobel Prizes, one might quickly reach the conclusion that the global center of scientific creativity is the USA (Fig. 2.19).

Of all the Nobel Prizes in science (i.e. physics, chemistry, biology and medicine) 34 % were awarded to researchers with a US American address (not all were born in America of course). After World War II, this percentage is 45 %.

That scientists from American institutes are so successful in terms of Nobel Prizes should not really come as a surprise. Evidently American universities and research institutes provide – on average – the best conditions for innovative research. Although school students in western Europe and Japan tend to outperform their American contemporaries, the Americans take the lead again at university level.

One should remember, however, that America is a country that has always attracted immigrants. Outstanding researchers from throughout the world have recognized that the conditions in their own countries cannot compete with those offered in America. Hence many have moved there, initially temporarily, but have then stayed and even become US citizens and/or Nobel Prize winners.

A somewhat different picture emerges, however, if one considers the number of science Nobel Prizes awarded per million inhabitants (Fig. 2.20). Here, surely, we have a way of quantifying national scientific creativity. Not really, for this statistic suffers from the fact that the samples are too small: A South-Sea island with one

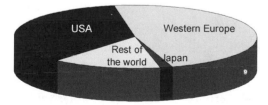

Fig. 2.19. Regional distribution of all Nobel Prizes 1901–2003

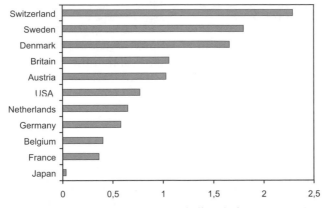

Fig. 2.20. National distribution of all Nobel prizes in sciences per million inhabitants (1901–2003)

Nobel Prize winner might appear to outperform the larger nations. The only vague conclusion that one can draw from this figure is that Switzerland, Sweden and Denmark seem to be home to an overproportional number of Nobel Prize winners. One may be tempted to conclude that the Nobel committee, based in Stockholm, shows an element of favoritism. Whilst this may have been true in the early decades of the Nobel Prize (local communication being that much simpler perhaps), it is surely not the case today.

There appears to be no direct correlation between the number of Nobel Prizes and the publication activities of a country or region. Despite the prolific number of publications emerging from Japan, only very few Nobel prizes have been awarded to Japanese scientists.

2.7.5 Patents

In principle ideas that are worthy of patenting should be strongly linked with creativity, since inventiveness is part and parcel of creating. However, it is not possible to patent a mere idea, nor a law of Nature (see Chap. 7 on Patents), and so patent statistics actually give a picture of a country's activity in technology rather than in science. But new technology relies frequently on prior scientific work in the same company or institute. Thus the statistics shown in Figs. 2.21 and 2.22 can also be considered as some measure of scientific creativity. Figure 2.22 shows the number of patent applications submitted to the European Patent Office according to nation of origin. To make a true international comparison, one would need to add the figures for submissions at least to the US and Japanese patent offices.

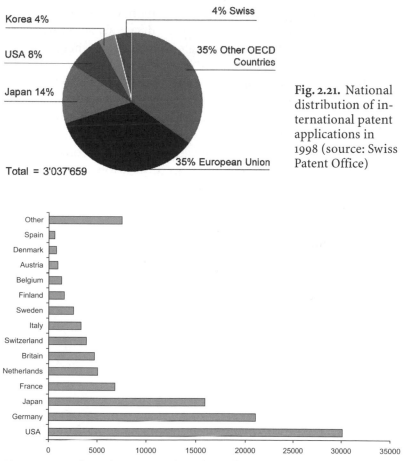

Fig. 2.21. National distribution of international patent applications in 1998 (source: Swiss Patent Office)

Fig. 2.22. Number of patent applications submitted to the European Patent Office according to applicants' place of residence or business (2002)

2.7.6 Conclusions

Our collection of statistics has gone a long way to confirming what we said at the beginning of Sect. 2.7: It is very difficult to quantify scientific creativity. We have succeeded, to some extent, in demonstrating regional differences in scientific output, but have found that, here too, different measures of productivity can yield conflicting conclusions.

Thus, as a young scientist embarking upon a career in research, our advice to you would be to base your decisions about where to work on more local factors, relating to the institute where you will work and the town where you will live. Who will be your immediate colleagues and advisors? Are these people known to be clever and innovative? Will you be able to research in an active and exciting area? Does the research program have enough money to spend? Do you feel at home in the environment provided? If you can answer most of these questions in the affirmative, then you can anticipate creative and enjoyable work, be it at a top US address or tucked away on a little-known island.

3 Scientific Presenting

'At a funeral, the average person would rather be in the casket than giving the eulogy.'

Jerry Seinfeld on the fear of public speaking,
quoted in *Time Magazine*, July 26, 2004

Most scientists, whether graduate students or experienced researchers, are called upon to make intermittent, if not regular, presentations. Indeed, in the course of their careers scientists usually give lectures or presentations more frequently than they write papers.

As we all know, excellent scientists are sometimes dreadful presenters. Helmholtz, for example, although a brilliant scientist, was notorious for his terrible lectures. Being a good scientist is certainly no guarantee that you will have a natural talent for presenting and, at the beginning of their careers, many scientists struggle with this task. A lucky few do find that it comes naturally; others learn quickly from their mistakes; but many more continue to give incomprehensible talks for as long as they remain in science.

The success of any presentation depends on three main factors: (1) the content, (2) the visual aids, and (3) the style of delivery. Whilst the preparation of good content relies on many of the same criteria and skills needed to write a good paper, the other factors – visual aids and style of delivery – involve completely different skills. This chapter will provide recommendations designed to optimize both the preparation and delivery of a scientific presentation.

Our aim is thus twofold: First, to draw your attention to numerous pitfalls that are often not recognized by presenters (whether students or experienced scientists) but are mercilessly criticised by audiences; and second, to give you the necessary information to prepare, practice, and perfect your presentations.

At whatever stage you make your first attempts to present the results of your research, if you do it properly, you can significantly speed up the process of establishing yourself as a recognized contributor to your field. And you can also save yourself much wasted effort and criticism later on. Your future students will also thank you; not only for the privilege of attending your clear and informative lectures, but also for the good example and advice that you will be able to pass on to them.

Before launching into the details, let us first glance at the various kinds of scientific presentation:

▷ conference papers

▷ seminar talks

▷ lectures to students

▷ defence of thesis

▷ defence of project

▷ poster presentations.

These different kinds of presentations all require exactly the same presenting skills. In terms of visual aids there is also not much to choose between them. They differ markedly, however, in the selection of material and level of detail that is appropriate. Not only do the lengths of presentations vary from 10 minutes for a short conference contribution to one hour or more for a seminar, but the audiences, too, will vary in their level of expertise and should thus also influence the content significantly.

In the following sections, where not otherwise stated, the advice will pertain mainly to short talks (typically 15–30 minutes) to be given at conferences. For most graduate students and young researchers, this type of presentation is not only most commonly occurring, but also – for many – the most daunting. Furthermore it is an occasion on which the speaker is critically judged by other, external, scientists (including potential future employers!).

3.1 Planning a Presentation

3.1.1 When to Begin

When you know that you have been assigned a conference talk or are scheduled to give a department seminar, you will immediately begin to think about what material you want to present. Whilst it is not sensible to worry about your talk for months in advance, it is good to prepare it fairly early, especially if you are relatively inexperienced. You will need time not only to collect and organize the material, plan the visual and oral content, and prepare the slides: for your first few talks, you should also allow a couple of days for practicing the talk (Sect. 3.4).

An acquaintance of one of the present authors (CA) gave an excellent presentation at a big conference. When I congratulated him on what, to my mind, was the best plenary talk (even though two others were given by Nobel laureates), he admitted that he had worked very hard on it for the past two weeks. This particular scientist was not exactly inexperienced and later the same year was himself joint recipient of the Nobel prize. This goes to show that even the best scientists have to put significant effort into producing convincing presentations.

As a rule of thumb, we would suggest that you begin preparing a short talk about two weeks in advance. For a longer presentation, three weeks is more appropriate. But of course a lot depends on how well you know the material.

3.1.2 Audience

Before selecting the information that you wish to present it is essential that you first consider the audience, something that many novices fail to do. Are you expecting a qualified audience consisting of specialists in your topic? This will be the case at a small workshop with narrowly defined topic, in a similarly specialised parallel session of a larger conference, or at an internal group sem-

Fig. 3.1. The audience assembles. Be sure to know their interests and level of prior knowledge

inar. Or will your listeners be students or scientists with a broad range of backgrounds interested in learning about a new field? If so, how much will they already know about your subject? And to what level of detail is it likely to interest them?

For specialists it is obviously wise to include more details, possibly sacrificing on breadth of information if necessary; but for a more general audience, it is appropriate to concentrate on the basics and describe the main results and most interesting phenomena, remembering to avoid, or at least explain, any technical jargon. Even an impeccably prepared talk will receive a poor resonance if it does not coincide with the audience's level of interest and knowledge. If you are faced with the problem of addressing a mixed audience of experts and non-experts, the best solution is to give your talk the same basic structure as you would for nonspecialists, but to add some more detailed and technical sections for the experts. This should allow everyone to go away feeling that they have profited from your presentation.

You might also ask yourself what will motivate the audience to attend your talk. Do they need the information for their own work? Is it necessary for their studies and exams? Or will they be attending purely out of interest for the topic and in the hope of learning some fascinating science from an enjoyable presentation. If you can answer this question, it will futher help you to avoid disappointing your audience.

As a rule, a talk will be best understood and most highly praised if most of the audience understand most of what you are saying. They may even have heard a good deal of it before. As remarked by Enrico Fermi one should *never underestimate the joy people derive from hearing something they already know*.

On one occasion I (CA) was invited to give a talk at the Japanese Conference on Applied Physics, based on an abstract that I had submitted. Unfortunately, some of the results that I was hoping to obtain before the conference were not forthcoming and I was forced to change the topic at the last minute. I decided instead to give a basic introduction to the new experimental method that was being used by our group. The talk contained virtually no new results, just an easy to understand overview of the technique, its principles and potential applications. Surprisingly it was greeted with great enthusiasm. I was thanked by colleagues for demystifying a topic that they had been struggling to understand. For me this was a valuable lesson in selecting the appropriate material and level for presentations. A misconception of many young students is that an audience of other scientists will already know a great deal about the research being reported. This is rarely the case: these days few scientists can acquire a detailed knowledge of more than one or two fields or of more than a handful of experimental techniques. Thus, as the speaker, you nearly always have the advantage of knowing more than your audience. Use this to instruct and inform your listeners, not to mystify and amaze them with incomprehensible statements.

3.1.3 Title and Abstract

For a conference contribution, the title of your talk will generally have to be chosen well in advance. It is then submitted, together with a short abstract, to the conference organizers in response to their call for papers. So even at this early stage you should consider the two factors (i) time available and (ii) audience expertise, when deciding how broad or narrow to make your title, and how technical to make the abstract. The same may well apply for other types of presentation too: For external and internal seminars or lectures, you will often be asked to provide the title and abstract of your presentation in advance so that it can be announced to interested parties.

The title should be made as short and as catchy as possible consistent with giving an accurate description of the material to be presented (see also Sect. 5.3.1 on choosing the title of a paper; for a talk, the criteria are the same). Note that the title is particularly important for a talk at a big conference with many parallel sessions. People will only come to your talk if its title promises that it will be interesting and informative. If you are working on a hot topic or have some exciting new results, don't be too modest to make this clear in both title and abstract.

Of course, it is not uncommon for presenters to deviate somewhat from the title announced in a conference book of abstracts, but it is not advisable to make a habit of it. And you should certainly not alter your topic so much as to find yourself speaking about chalk in a session on cheese.

When eminent scientists are invited to give plenary lectures at conferences they are usually free – within certain limits – to specify the topic and title of their talk. Some time ago, I (CA) met the Nobel laureate Gustav Hertz, already in his mid-eighties, after one of his presentations. I complimented him on his talk and on the fact that he was still actively preparing and delivering scientific presentations. His response: '*Young friend, for years now I have always given the same presentation. The conference organizer simply asks me to change the title a little and occasionally the emphasis.*' Unfortunately, as a newcomer you cannot afford this luxury.

3.1.4 Collecting Material

If you will be speaking to an audience of non-specialists, you will have chosen a fairly broad topic. The information you want to deliver will then focus on the essential ideas and overall importance of the field. For a specialist audience, on the other hand, you will probably wish to present much the same type of information as you would publish in a short paper.

With these prerequisites in mind, it is often useful, before beginning to structure your talk, to simply make a list of ideas, in no special order, of points that need to be mentioned, either on a slide or orally. To facilitate the later selection of key points for slides, it might be useful to classify these points into '*musts*' and '*maybes*'. Sometimes these points will fall naturally into a logical order, but more often you will have to impose the detailed formal structure later on, adding and deleting material as you go along. This is also the time to collect all diagrams and other illustrations that you may want to use.

During the process of collecting information, it is advisable to keep in mind the following points:

1) What is the state of the art in the field? If possible this should be summarized first in order to set the scene for your results and justify the importance of your research project. If confirmatory or contradictory results have been obtained by other researchers be sure to know about these. If not in the talk itself, you may need this information for the discussion.

2) What supportive information is needed for the audience to gain a full appreciation of the results? E.g. some basic physics, a little history, some experimental details. Such additional or background information can make a talk more interesting and takes advantage of the unique opportunity to attract attention to your research. What are the main results to be presented? Can you present all these in the time available? If not, reduce the list from the outset.

3) Unless you are preparing a plenary or review talk, which may, but need not necessarily, contain new results, make sure that you have at least some novel information in your presentation. Specialists will be disappointed if they go away having learned nothing new.

In parallel with – or, if you prefer, in advance of – this process of compiling and sorting material, you should begin to develop an overall plan for your presentation. Make a note of this backbone of key elements and use it to begin ordering your material as soon as you are confident to do so. The order of points should be both logical and didactic, allowing the audience to follow the argument effortlessly. Examples of possible backbone structures are:

For a Short Talk

▷ Introduction/state of the art

▷ The unsolved puzzle

▷ Experiment to shed light on puzzle

▷ Result

▷ Interpretation

▷ Open question(s) for further work

▷ Conclusions/Summary

For a Longer Invited or Plenary Review

▷ What is the XYZ effect?

▷ Why is it important and interesting?

▷ What open questions are currently being tackled?

▷ My contribution to the mosaic

▷ Summary of the main results so far

▷ What overall picture is emerging?

▷ Outlook for the future

The purpose of such a structure is not to define the final content of your talk. It is simply useful as an intermediate step, helping you to organize your material and your thoughts into a coherent story with a beginning, middle and end. Every talk, however long or short and however general or special should have the nature of a story. As we know, some stories have a happy end, other equally good ones leave the reader with a challenge to take away and ponder. So don't worry if your talk does not give the definitive solution to the problem.

If the title has not already been decided in advance, now, having decided on the overall scope of your talk, you may like to devise a good title, which in itself can guide you in further structuring your material and finding the best emphasis.

With these preparatory steps completed, you are now ready to put the flesh on the bones.

3.1.5 Detailed Structure and Content

The rough backbone structure used to help select and order your material is probably not sufficiently detailed to completely define the *'finestructure'* of your talk. Here it is useful – for a conference contribution at least – to adhere roughly to the same logical structure required in scientific papers (see Sect. 5.3), again keeping in mind that the overall story must be logical and didactical.

However, if a presentation were to be given simply by reading out the text of a research paper the audience would barely be able to follow it. In an oral presentation the information should not be packed so densely. When absorbing information from a written paper, the reader can stop and think about the statements just read; can re-read them; can go away and check another source in order to clarify the meaning. This is not possible during a talk, and implies that an oral presentation should use significantly more words than a paper to explain the same idea. Thus in an oral presentation you should restrict yourself to the main facts and the main steps of an argument. In between you should give the audience time to absorb the information, perhaps by rephrasing it in another way or filling in a few details to support the claim, or indeed simply by speaking slowly and pausing every now and then.

The didactical element is also more important in a talk than in a paper, even when you are speaking to a specialist audience. Thus, you should aim to use plenty of analogies, simplified arguments and appealing ways of visualizing information. If you need to explain ideas that are new to the audience, be sure to do so in terms they can follow, avoiding technical jargon. Keep the science as simple as possible but the content as interesting as possible. You want to ensure that your audience leaves with the feeling that they understood everything and learned a lot.

The next step in organizing your material should be to decide how many slides/transparencies you will show and to divide the material up into appropriate chunks, making each as self-contained as possible so that every slide shows one chapter of the story. For PowerPoint slides, it is advisable to plan at least one-and-a-half minutes per slide, meaning that you can show 10-15 slides in a typical short conference contribution. If using conventional transparencies with an overhead projector, the number should be reduced slightly to allow time for handling (and mislaying!) the material.

The order of the slides will correspond more or less to the backbone structure already devised, but some of the key points will need more than one slide to do them justice. In Sect. 3.3 on preparing slides we will discuss in detail the question of how much or how little information to show on each slide. In the remainder of this section we address the matter of defining the full spoken content of the presentation.

It is not necessary or desirable to write out the text of your talk longhand (the slides themselves will help you to remember the main points), but it should be carefully thought through and planned, and, for practice purposes, beginners may even wish to make more extensive notes about the content than those appearing on the slides.

As with a scientific paper, it is sensible – when reporting on new results at least – to divide the talk into several or all of the following parts:

Introduction – What's the Problem?

Your very first words will probably be something along the lines of *'Today I would like to tell you about our work on ...'*, at which point you read out the title of the talk. This is a safe, if rather undramatic, way to begin a presentation. At the start of a longer talk, it is often useful to give a very brief outline of what is to follow (but a luxury one can ill-afford in a 10-15 minute talk). Thereafter, the topic is briefly described, together with the motivation for the research to be discussed and the questions that you are trying to answer.

It is essential to gain the interest of your audience during the introduction: Your first two or three sentences are often enough to shape the audience's attitude to your talk as a whole. Once you have lost their attentiveness in a longwinded, boring, confused, or inaudible introduction, it will be very hard to retrieve it later! Thus think very carefully about the content of the first few sentences, and practice their delivery carefully (see Sect. 3.4). If you are confident enough to use some humor here and have an appropriate opportunity, this can help. A funny quotation on the first (title) slide is probably the easiest way to introduce a light-hearted element. But if you don't feel comfortable with humor and are afraid of it missing the mark, then it is better to make *'clarity'* your top priority. Your introductory remarks must convince the audience that what you have to say is important and that you yourself are alert and competent and – not least – that you believe in what you are saying.

Experiment and Method – What Are We Doing?

This section, usually brief, should outline the methods and analytical techniques used to acquire your results. If you have developed your own experimental method or novel procedures these should be clearly described. To forestall questions, you might usefully mention sample preparation and data processing methods, if there is time. Even if they do not warrant explicit mention, make sure you can provide the details if asked.

Results and Discussion –
What We Have Found and What It Means

As for a written paper, this part usually forms the main body of a presentation. In written reports, the results and discussion are sometimes divided into two separate sections. In a talk this is not advisable. The results should be discussed immediately after they are presented and while they are still fresh in the minds of the listeners. Another difference: In a paper, except perhaps a letter, there is sufficient space to include numerous sets of detailed data together with a lengthy interpretation. For a short talk you probably won't have time to include so many results and a careful interpretation of all their features. In any case, this would almost certainly overload the audience with information. Thus it is important to select the most significant results and to present these clearly and convincingly. When results are presented in graphical or tabular form on a slide, it is essential that you also describe these curves and figures in words – the audience should need to *think* as little as possible but to *understand* and *absorb* as much as possible.

When discussing your results it is important to refer to related results of other researchers. But these should not be limited to results that confirm your interpretation: any contradictory results should also be mentioned, if possible with an explanation to explain the discrepancy. Don't forget to stress what is new about your results.

Give a clear interpretation of your results as far as this is possible. But rather than drifting off into vague speculation and further incomplete interpretations it is better, once the firm interpretation has been given, to describe possible future experiments which would answer any open questions and confirm uncertain hypotheses.

Conclusion – Summary and Take-Home Message

It is common practice to give a summary at the end of a presentation. Here the main points of the talk are repeated in a concise list without supplying any new information This will often involve re-emphasizing, and perhaps re-wording, the key points from the discussion. An alternative and probably better way of finishing a presentation is to draw some general conclusions. Often it is sufficient to simply list the main scientific conclusions and implications of your work, without recapping on how these were obtained. You should also briefly discuss the wider implications of your results, assuming any are already recognizable. But the important thing to remember when choosing your concluding words is that the audience should be given a clear and simple message to take home. Since most people cannot remember more than three things at once, try to limit your conclusions to no more than three major points. Then your talk will be remembered as one of the highlights of the session, if not of the conference.

Acknowledgement – Who Did and Paid for the Work?

It is good practice to acknowledge those people and organizations who, besides the named authors, have made significant contributions to the work presented. These may include collaborators, technicians, discussion partners, head of department, but also the funding agencies and any other source of grants. For invited and plenary talks, which at some conferences are only allowed to have one author, it may even be necessary to acknowledge co-authors who would otherwise go unmentioned.

The acknowledgments are often given on the final slide of a presentation. Many speakers leave this slide visible at the start of the discussion. If you choose this option, it is not essential that you read out the content of the slide unless you wish to stress particular gratitude to some party.

Another means of acknowledging support – and the preferred choice in our view – is to include the details on the first slide beneath the title of your talk. The advantage of this is that the acknowledgements do not distract the listener from the message you are trying to imprint in his/her mind at the end of your talk. If there are no acknowledgements following the take-home message, the latter can be left visible during the discussion, surely a good way to increase its impact.

> **And finally …** How **not** to structure your talk (although we all know of scientists and lecturers who use this method!):
>
> ▹ In part I, the audience hears things that they already know to make them feel secure and to show them that they are not so ignorant.
>
> ▹ In part II, they hear things that are new to them so they can feel that they are profiting from the talk.
>
> ▹ In part III, they hear things that are completely incomprehensible to them to make sure that they remain in awe of the speaker.

3.2 Visual Aids

3.2.1 Preamble

A scientific presentation is not a political speech, nor is it a radio broadcast, and still less a sermon. Words are not enough.

Visualization of the information is essential and the quality of the images can be a make-or-break factor for the success of a presentation. As the old saying goes: A picture is worth a thousand

Fig. 3.2. Preaching (Wilhelm Busch) … but in a scientific presentation words are not enough

words. This is especially true when it comes to conveying complex scientific and technical information. For example, it is much more effective to present your data in the form of simple plots than to include tables of numbers or lengthy oral descriptions of the results. Indeed, many trends and dependencies only become recognizable when represented in graphical form. When speaking, too, you should aim to make use of pictorial language and visual analogies.

Some people learn more effectively by seeing/reading, others by listening to words. The oral description coupled with the visual representation allows you to exploit these two sensory channels of communication simultaneously, thereby strengthening the impact. Even the sense of touch can be utilized in some talks, for example by passing samples around the auditorium. Additional visual impact can be gained by doing simple experimental demonstrations. A very famous example was Richard Feynman's demonstration, performed at the inquiry into the first challenger disaster, of how O-rings become brittle at low temperatures.

For teaching purposes, too, a demonstration can often be very effective. This is clearly recognized in the remark, attributed to Jearl Walker of Cleveland State University, '*The way to capture a student's attention is with a demonstration where there is a real possibility the teacher may die.*'

The majority of talks, however, rely exclusively on visual information provided by means of an image cast onto a screen or wall behind the speaker. This image can be produced in at least three ways: on a computer/laptop, on a conventional transparency, or on a photographic slide. In the following sections, we will discuss all of these techniques briefly. But we note from the start that computer presentations have – for good reason – now become by far the most popular. In another few years it is likely that transparencies and photographic slides will have virtually vanished.

3.2.2 Computer Presentation with a Projector

With this technique, the slides appearing on the monitor of a PC or laptop are cast onto a wall or screen using a video projector. For

Windows-based computers, the slides are typically produced using Microsoft PowerPoint, a standard component of the Office software suite. This is also available for Macintosh computers, but Mac fans may prefer to use Apple's own presentation program, Keynote. In the remainder of this section, all that is said about PowerPoint applies equally to Keynote and other similar presentation software.

Computer-based presentations have several advantages over the use of old-style transparencies, but they also harbour a number of pitfalls.

The programs are relatively simple and intuitive to use. If you take advantage of an existing layout template, or create your own, the slides automatically have a uniform and easy-to-read style. Once prepared they are technically simple to deliver; rather than fumbling around with loose sheets of paper, you need only press one key to move to the next slide.

Secondly, the information can be built up step by step as you speak. This ensures that the audience is not distracted from the point you are making by something that is to follow. It is particularly useful for complex figures, such as graphs showing several curves. By adding the curves one at a time, explaining the differences as you go, you will leave the audience with a much clearer idea of the underlying relationships.

The slides themselves, in contrast to conventional transparencies, are wider than they are long. This format helps to overcomes one of the problems that was previously often encountered in larger lecture theatres, namely, that those in the distant parts of the audience could not read the lower lines of a transparency.

PowerPoint also offers the possibility to include animation: text or images can be made to slide into the picture from the left, right, above or below, or to fade in and out of view. But herein lies one of the potential pitfalls. Whilst such animation may help to keep the attention of the audience, you must be aware that excessive use of gimmicks can also distract from the real message. So they should be used sparingly or not at all (see also Sect. 3.3).

Colour can be used freely, and the number of colours available is limited only by the monitor settings. But, as with the animations, overuse of different colours can become a distraction rather than an advantage. A rainbow coloured text contains no more information than a black and white text, and will be far less effective at conveying this information.

Before preparing your presentation on the PC, you should satisfy yourself that appropriate and compatible facilities will be available in the auditorium. Even if you do not expect any problems, it is wise to take a corresponding set of conventional transparencies with you. And, although you may be intending to use your own laptop for the presentation, it is advisable to take with you a CD containing your talk so that you can deliver it from another computer if necessary.

Sometimes it is expensive to use a projector supplied at a conference. At a recent meeting of the American Physical Society, speakers had to hire the video projector from the convention centre at a cost of $ 200 for a 20-minute presentation. Use of the overhead projector, however, was free.

A final advantage of PC-based presentations is the ease with which you can modify a talk – even five minutes before you are due to speak – by adding new slides or correcting existing ones. If there is good reason to do so, you can even add references to the presentation that preceded your own.

3.2.3 Conventional Transparencies

Until recently these were a very common way of preparing and delivering the visual information that accompanies a presentation. They are simple to create and transport, require no computer skills, and involve no risk of compatibility problems. Moreover, appropriate equipment for projecting transparencies is still to be found in virtually every lecture theatre in the world. And you do not need to transport extra weight in the form of a laptop.

Nowadays, however, it is mainly the older generation of scientists who favour this technique of presenting. There is no doubt that visual aids produced on a PC are generally of a much higher quality that the corresponding '*overheads*', which are often hand-written, smudged, or out of focus. Thus by opting for conventional transparencies you may be doing your audience a disservice. Worse still, you are probably doing yourself a disservice by creating the impression that you are not computer literate, technophobic, or simply out of date. For job applications in particular, this will hardly count in your favour.

3.2.4 Photographic Slides

In the 1960s and 1970s it was quite common for talks to be accompanied only by photographic slides, which were projected onto the screen by a slide projector at the back of the auditorium. By the 1980s most presenters were making use of an overhead projector with transparencies. But slides continued to be interspersed when high resolution photographic images needed to be shown. Since the advent of scanners and electronic photography, one rarely encounters this method of projection. Its only advantage is that it still offers a much better resolution than images projected from a computer monitor.

3.3 Preparing Slides

Poor preparation of visual material is one of the main reasons why talks fail to please their audience. The following guidelines – which, for the most part, apply equally to the preparation of PowerPoint slides and conventional overhead transparencies – should help you to optimize your visual aids. Note, it is not sensible to start preparing your visual aids until the detailed content of the talk has been fixed, if not on paper, then at least in your mind.

3.3.1 How Many Slides?

Depending on the time available, estimate in advance how many slides you should show. As a rule of thumb, you should allow at least one-and-a-half to two minutes per slide. This is simply the amount of time the audience will need to read and digest the text of a typical slide. If you know that many spoken words will be needed to elucidate the information appearing on the slide, you may need to allow considerably longer. This means that, for a short conference presentation (ca. 15 minutes), the optimal number of slides will be 8-10.

Then divide your material into this many units, not forgetting an introductory and concluding slide. If the final slide is an acknowledgment that will not be read out, you need plan no extra time for it. Try to devote one slide to each main topic or part of the argument. Note, however, that it is better to have one or two additional slides than to squash too much material onto each. With the possible exception of self-explanatory jokes or cartoons, all slides should be conceived in such a way that they will correspond to at least one minute of oral explanation.

3.3.2 Amount of Information

Perhaps the most important piece of advice that can be given here is: Do not overload your slides. The smaller the amount of information on a slide, the greater its impact will be. Suppose a lecturer wants to convince her audience that a certain hypothesis is untrue. More effective than a slide containing ten equations to prove the impossibility of the hypothesis is one containing nothing but the statement '*NO!*'. Whilst displaying this, a few words explaining how this conclusion is reached can be substituted for the detailed equations. This is an extreme example and there will of course be occasions when some equations are necessary. But the message '*less is often more*' should always be kept in mind as you prepare your visual material.

For a conventional overhead transparency, a golden rule is that it should contain no more than 13 lines of text if legibility is to be maintained. For computer presentations, on the other hand, 13 lines of text may already be too much. Here there is another general rule that says: 30 words per slide are enough. This can correspond to anything between three and eight lines of text, although we would recommend a maximum of five points.

3.3.3 Headings

The best structure to aim for consists of one main heading on each slide followed by a few lines containing relevant and concise statement or keywords. The headings should be chosen to be short and accurate but catchy, so as to provoke the audience's interest. Often it is a good idea to use a short question as a heading. For example, having clearly explained your new hypothesis in the introduction, you may like to entitle your next slide '*How Can We Test This?*' – and then go on to outline the experiments you have performed. But like all good ideas, this one should not be overused. If in doubt, a dull but accurate heading is better than a catchy but misleading one.

3.3.4 Keywords and Key Statements

The body of the majority of slides will consist of a list of brief phrases. You should not formulate these as full sentences and then simply read them out. The audience will already have read them and, if you supply no information beyond that on the slide, your reading, indeed your presence, will become superfluous.

On the other hand, the keywords or key statements should be descriptive enough to convey a message. Rather than just writing '*Cell division*' as a keyword, which will undoubtedly leave the audience guessing, it is better to include some relevant information: '*Decreased cell division after two days*'.

Thus the list of key statements serves the dual purpose of providing the audience with the main facts and arguments whilst simultaneously reminding you, the presenter, of the essential points that you want to mention and elaborate on.

3.3.5 Typography and Font Sizes

Use different font sizes and/or different colours to distinguish between headings and other text. You may even want to introduce a third type size for sub-items on lists, but beware of overloading slides in this way. Also, be careful to keep to a consistent scheme for writing (and, where appropriate numbering) the headings or sub-headings. What font size is needed? This will depend on the size of the lecture hall. Be sure, however, that all the text is in a font large enough, at least 18 pt, to be read comfortably by those in the back row of the audience.

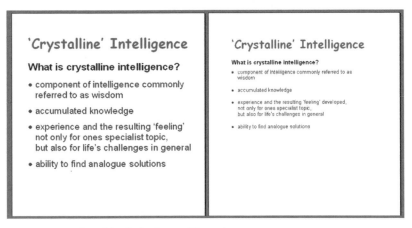

Fig. 3.3. Good and bad choices of font sizes

Different fonts are also useful for placing emphasis. Unlike in a written report, where italic is recommended for emphasis, on a slide it is probably better to use boldface. Why? A written text is usually read from start to finish and the reader will interpret the italic as emphasis when reaching it. In this case bold text would prematurely attract the eye away from the rest of the text. On a slide, however, you may want to do exactly this: attract the viewers' eyes to the one phrase above all others. For this purpose boldface is the most effective.

3.3.6 Diagrams

Diagrams and other images must also be legible. That means, in particular, that all lines should be strong enough and dark enough to be clearly visible. Like the text, the figures should be restricted to the main information. Do not include all 36 stages of the apparatus for cleaning and processing semiconductor wafers. Whilst such a figure can be useful in a publication, where it can be viewed at leisure, it is next to useless in a presentation where you only want to draw attention to one stage. Redraw or cut out the relevant part of the figure and show only this. On graphs, all axes and curves should be clearly labeled. Avoid using only symbols to label axes. Even though a scanned figure may not be clearly labelled, you can use PowerPoint to overlay suitable words. Photos or any other images that are not absolutely self-explanatory should include a concise one-line description.

3.3.7 Mathematics

The material displayed on the slides should be restricted to information that is either well known to the audience present or can be readily followed on the basis of the talk itself. Thus complicated mathematical derivations that need a long time to assimilate should be avoided. If a mathematical derivation is an essential part of the work being presented, it is better to summarize the steps in words and just give the major results.

There are of course those who cannot resist the temptation to assert their mathematical superiority. They mistakenly believe that their professors and peers will be impressed by expressions that look as complicated as possible; and that the elegance of the equation

$$\ln(e) + \sin^2 x + \cos^2 x = \sum_{n=0}^{\inf} 2^{-n}$$

far outweighs the convenience of its more trivial form

$$1 + 1 = 2.$$

Your audience, however, will thank you more for applying the KISS principle: *Keep It Short and Simple.*

$${}^t q'_{13} : T_{(y_o, p_x, p_z^a)}(Y \times T^*(X \times Z)) \hookrightarrow T_{(p_x, y_o, p_z^a)} T^*(X \times Y \times Z)$$

is transversal to $\mathrm{Im}(\delta_\pi^{-1}(\lambda_1 \times \lambda_2) \xrightarrow{\imath \delta} T_{(p_x, y_o, p_z^a)} T^*(X \times Y \times Z))$, where $y_o = \pi(p_Y) \in T_Y^* Y$. Now, let $\rho' \subset \rho \subset T_{(p_Y^a, p_Y)}(T^*(Y \times Y))$ be the isotropic subspaces defined by

$$\rho'^\perp = T_{(p_Y^a, p_Y)}(Y \times_{Y \times Y} T^*(Y \times Y)) \text{ and } \rho = T_{(p_Y^a, p_Y)}(T_Y^*(Y \times Y)) \ .$$

Then we have:

$$T_{(p_x, p_Y^a, p_Y, p_Z^a)} \left(X \times Y \times Z \underset{X \times Y \times Y \times Z}{\times} T^*(X \times Y \times Y \times Z) \right) \simeq E_Y \oplus \rho'^\perp \oplus E_Z^a \ ,$$

$$T_{(p_x, y_o, p_Z^a)} T^*(X \times Y \times Z) \simeq (E_X \oplus E_Y^a \oplus E_Y \oplus E_Z^a)^{(0 \oplus \rho' \oplus 0)} \ ,$$

and,

$$T_{(y_o, p_x, p_Z^a)}(Y \times T^*(X \times Z)) \simeq (E_X \oplus \rho \oplus E_Z^a)^{(0 \oplus \rho' \oplus 0)} \ .$$

Fig. 3.4. Mathematical overkill

Are Perfect Slides Always the Best Way? For long and complex mathematical formulas, a slow step-by-step derivation using blackboard and chalk may well be easier to understand than a slide containing the same material displayed as a '*fait accompli*'.

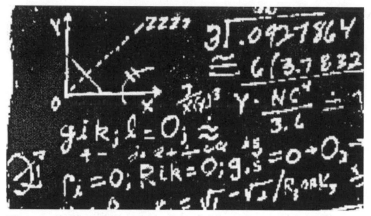

Fig. 3.5. Chalk and blackboard: useful for extra explanation

3.3.8 Use of Colour

Your slides should be pleasing to the eye and should naturally guide the viewer to the most relevant information. Therefore, colour should definitely be used, but sparingly. It must fulfill a purpose. For example, you should use colour to help structure the information, for headings, and perhaps for emphasis. But not for decoration: A rainbow-coloured text contains no more information than its black and white counterpart. And it will distract the viewer instead of helping to transmit the content.

The same considerations apply to the use of colour on figures. Here, especially, many presenters make the mistake of using colours that are too light. Deep shades of red, blue or green and black are the best choices and should provide sufficient for most purposes. If you need to use many colours, for example for a pie chart with many elements, take care to place contrasting colours next to each other and to provide a key in which the order of the colours corresponds to the sequence in the diagram. Avoid clashing colours such as red text on a green background since these are strenuous to read and distract attention from the content. Recall too that a significant fraction of people suffer from red-green colour-blindness.

3.3.9 Revealing Information Step-by-Step

When using conventional transparencies, some presenters have the habit of covering part of their transparency while discussing the first point and then gradually revealing the subsequent points by sliding the coversheet down. This forces the audience to concentrate on what is being said rather than reading on ahead to see how the story proceeds. Whilst this has its advantages, it can be distracting for the audience and is perceived by some as a restriction on their right to choose what they want to read. It is probably better to limit the amount of information from the start. PowerPoint, however, has a built-in facility for revealing the information one text field at a time. Most people are now accustomed to this and do not object to its use. But do you, the speaker, remember what is on the next line? If in doubt, it is better to display all the blocks of information at once.

3.3.10 Animation

In addition to the simple building up of a slide step-by-step, Powerpoint offers many other features allowing one to make text slide into the picture from all directions, one letter or word at a time; to fade in or out of view; to rotate about an axis; to flash on and off, … Pictures and diagrams can also be juggled in the same way. As we argued concerning the use of colour, these features should only be used where they serve a purpose (humour is a valid purpose of course!). If you overdo the use of these features you will find that the audience is mesmerized by your whiz-bang special effects, but has absorbed nothing of the message that you actually wanted to convey.

A case in point: My (CA) daughter, who is a student, had to give a presentation at a seminar. She prepared this very carefully and fully exploited the opportunities offered by PowerPoint: Every slide was built up in a different way, the text appeared from the left, then the right, from the centre or in stripes. Sentences appeared word by word, key statements one letter at a time. One slide faded into the next and a hammering sound accompanied the points that she wanted to hammer into the listeners' memory. The figures were built up one element at a time into final pictures that were as colourful as any Picasso. Her professor's comment at the end: *'The content was undoubtedly good, but I was unable to appreciate it because my attention was fully occupied by the impressive animation!'*

So let us repeat: The use of animation should be restricted to simple effects that support the content and do not become independent entertainment. Occasionally it is good to include a cartoon or some *'clip art'*, a large selection of which can be found on the web, but here too, it is important that the joke is one that helps to stress your message.

3.3.11 Handwritten Transparencies

Handwritten transparencies are to be avoided if possible. In the time needed to prepare a clear handwritten transparency, you can prepare two even better ones on the computer. If you have struggled to read transparencies such as the following example, you will not want to put your audience to the test.

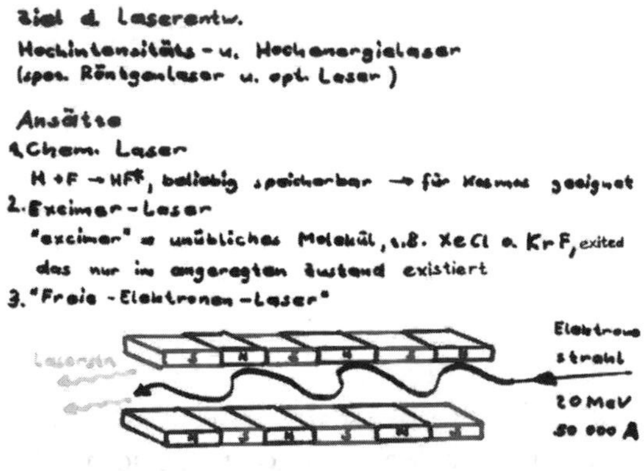

Fig. 3.6. Handwritten hieroglyphs

However, if – for whatever reason – handwritten transparencies are your only option, you should simply adhere to the advice given above concerning the amount of information, size of writing, and use of colour. Then even handwritten transparencies can be clear and informative.

3.3.12 Check List

Before you leave the office for the meeting where you will deliver your talk, make sure that you have with you all the necessary items. These will include some or all of the following:

▷ laptop with saved powerpoint file

▷ any connection cables that you expect to need

▷ CD-ROM or diskette containing the same file as a back-up in case your laptop goes on strike or cannot be connected

▷ set of transparencies showing the same information

▷ printout(s) of your talk (useful for discussions afterwards)

▷ non-smudging pens and spare transparencies.

3.4 Practicing Before the Event

Having followed all the relevant advice in the previous sections, your talk should now be perfectly prepared. The right amount of content has been selected and organized and your visual aids (transparencies or PowerPoint slides) cannot be faulted. So far, so good. But this is only the first step. Carefully selected information and perfectly prepared slides/transparencies are not in themselves a guarantee of a good presentation. Just as important is the style of delivery. Thus, a few days before you are due to give your talk, it is wise – especially if you are inexperienced – to spend some time planning and practicing the actual delivery of the talk. Some people are born presenters and have no difficulty in capturing and holding the attention of their audience and in telling their story in a lively and convincing style. At the other extreme are those, including many experienced lecturers, who have little such talent and do not even make the effort to learn from their mistakes. All of us can remember the very varied quality of the lectures we attended as students. However brilliant a scientist he or she may be, a poor lecturer can ruin a whole course for the unfortunate students who suffer through their boring and incomprehensible lectures.

But even those of us who are not gifted as presenters can learn to master this challenge through careful preparation and practice. It may be useful to practice the talk on your own, but better still is to practice in front of colleagues and to ask for their comments and criticisms. Allow them also to cross-question you about the content of the talk. Perhaps you have missed out an important step in the argument or are contradicting yourself somehow. It is better if such weak points are discovered by your colleagues than in front of the hundred or so strangers you were hoping to impress. Your colleagues will also be able to identify questions likely to be asked after your talk and thus give you an opportunity to prepare convincing answers.

Even friends and relatives who do not understand the content of the talk may be able to give you some useful advice about your style of presenting (speed, audibility, perfomance, etc.). In Sect. 3.5.7, at the end of the detailed discussion of delivery, we give a checklist that you can give to members of your practice audience (or even listeners in a real presentation) so that they can evaluate your overall performance and identify the main strengths and weaknesses of your content and delivery.

If you have the opportunity, try recording your presentation on video as this can be particularly revealing. Get other people to watch the video with you and give comments. But you yourself should be able to identify most of the things that need improvement. One of the things that is likely to surprise you when you play back the video is your voice and style of speaking. Analyse this objectively. Then, if necessary, practise speaking more slowly and deliberately.

Fig. 3.7. A video recording of your talk can be very embarrassing ... er ... we mean illuminating

You can also learn a lot by analysing the presentations of others. In this respect, but only in this respect, you can learn far more from a bad presentation than a good one. It is much easier to identify the mistakes that make a presentation poor than the factors that give rise to a smooth, clear and convincing presentation. Try it and see!

It is indeed a shame that most of our university lecturers have no formal training in methods of teaching and presenting. But, like the young presenters primarily addressed here, even the *'old hands'* could – if so inclined – greatly improve their presenting skills by following our simple recipes. Once, I (CA) was sitting next to an eminent scientist, also a Nobel laureate, at a conference. After his talk I congratulated him on what I felt had been a virtually perfect presentation. He remarked *'But this was not always the case. I have learned a lot over the years from attending other good and bad presentations'.* This is an encouraging admission for us mortals: It demonstrates that even famous scientists may initally have had to struggle with their presentation technique. And perhaps one could go even further: Making the effort to perfect your talks can help to pave the way to the top (not only in science).

Anyone, therefore, who is not confident that they can give a good presentation should practice before each talk. We recommend such a dry run for your first three or four presentations. A further benefit of good preparation and plenty of practice is that you will be less nervous. Some other tips that will help to allay your fears are the following:

Are You Afraid of Forgetting Important Points? If you have pre-pared your slides well, then they should contain keywords that will remind you of all the important points you wish to make and each of the crucial steps in your argument. If you want to remind your-self of additional points not included in the slides, you could write some extra notes on a hand-held paper copy. But this is not really to be recommended as it means you will have to handle and look at these sheets, which will interfere with the smoothness of your pres-entation. For those using transparencies: Choose the type with a paper frame so that you can add notes on the frame to jog your memory. Remember though: you are the only one who knows all the things you planned to say. Even if you forget some points, this will be no great loss provided your talk is still coherent and con-vincing. It is better to forget 50% of the content and reach 100% of the audience than to remember 100% of the content and reach no-body.

Are You Worried About Exceeding the Time Limit? You can check the time needed quite accurately by giving the talk on your own. But in doing so you should include the handling of your laptop or transparencies and at least simulate the use of a projector – these things also take time. In the dry run, aim to take a little less than the allotted time; when speaking to an audience you may automat-ically slow down in order to let the information '*sink in*', a process that you can usually follow by observing the audience's faces. Final-ly, have one or two slides, towards the end of your talk, which you regard as non-essential. Then, depending on what the clock says, you can decide during the talk whether or not to discuss them.

Are You Afraid of Not Finding the Appropriate Words? This can indeed be a problem if you are new to presenting or if you have to give a talk in a language other than your native tongue. One possi-bility would be to write out the whole text of the talk and to learn it by heart. Provided you have learned it well, the audience will not notice that you are reciting from memory. But you will need plenty of time (ca. one week) and practice if using this method. In fact, learning by heart is not really to be recommended, even to beginners. If you lose the thread it might throw you completely off course and if you are not word-perfect, the audience is likely to notice the lack of spontaneity in your style. With time and experience you will become more at ease when presenting, will have a better grasp of the scientific content of your talk, and – provided the talk is carefully planned and prepared – should find that the right words come to you automatically. Whatever you do, do not succumb to the temptation to read the text from a script! Even a slightly erratic freely delivered presentation is better than listening to someone whose nose is buried in a sheaf of paper, who speaks in a monotone, and has no eye contact with the audience. For a statesman giving a political speech it is excusable to read from a prepared text, since a single wrong word can precipitate a political crisis. Scientists, fortunately, are usually more forgiving.

Finally, a note on repeat performances. If you are giving the same, or a similar talk for a second time this should not lull you into a false sense of security. Even after a short gap it is sensible to review the material and do at least one dry run.

3.5 Delivering a Talk

When delivering a presentation you are not only presenting its contents, but also yourself. This important aspect is overlooked by many speakers. The audience will form an opinion not only about the work you have done, but also about your competence, enthusiasm, and personality. Whilst these are not things that you can change for the sake of a presentation, many presenters succeed in hiding them behind a veil of confusion, inaudibility and/or dullness. The audience will appreciate you most if you give them a talk that is – in order of importance – clear, informative, interesting and, especially in a longer presentation, entertaining.

The people attending your talk are giving you a valuable commodity: their time. They do this in the expectation of being rewarded with stimulating and useful information. They are also ready and willing to be convinced of your story. But before you can convince the audience, you have to be convinced yourself. Hence you should be satisfied that your experiments, calculations, and conclusions are correct and sensible. When you are delivering a message that you believe in, you will automatically deliver it with more confidence and sincerity.

But sincerity, of course, is not enough. In the following sections we discuss in turn the additional key factors that contribute to the successful delivery of a talk or lecture. You should of course think about these matters well in advance and work on them during the practice sessions.

3.5.1 Style and Manner of Speaking

Free Speech. As was stressed earlier, it is essential that you speak freely rather than reading the text of your talk from a script. Reading – inevitably accompanied by a bowed head – means that you have little or no eye contact with your audience. You may as well be talking to a wall. Reading aloud also tends to come across in a more monotone voice, with the result that the audience's attention soon starts to wane. If you know what you want to say and have sufficient keywords on your transparencies, there should be no need to read anything from a script.

But free speaking in a lecture theatre is not the same as telling a joke to the person sitting next to you in a bar. It demands much more attention to clarity, speed and intonation. Your voice has to carry to the back row so it has to be loud and clear and your pronunciation of the words must be more careful.

Rhythm and Dramaturgy. The speed at which you speak should be slower than in a normal conversation. The slower speed is needed both to avoid acoustic jumbling of the words in a large room, and to give the listeners enough time to absorb the information, particularly those parts that are new to them. Once members of the audience have lost the track of your argument, you have lost them for

Fig. 3.8. Half the battle is won once you have mastered the art of free speaking with attention to eye contact and clear, unhurried articulation ...

the rest of the talk. Hence, as well as speaking slowly, it is often good to pause slightly between sentences and certainly at the end of each important statement. This is not only for dramatic effect; it also allows the foregoing message to be digested. Remember also to modulate your voice in order to stress the essential points. It may be useful to study the habits of successful speakers such as Bill Clinton. You will find that they speak with a well-defined and pronounced dynamics, varying their speed of speaking and, in particular, modulating their intensity to place emphasis. Do not underestimate the value of these aspects in shaping the message that is received.

As a scientist you are unlikely to have been trained in dramaturgy in the same way as an actor, a politician or a Roman senator. Fortunately, other scientists are tolerant in this respect, since most are themselves just as insecure of their theatrical and rhetorical abilities. So really there is nothing to fear. Knowing this, and having practiced hard, you should quickly gain the confidence necessary to give a convincing performance.

Fig. 3.9. ... but, on the other hand, don't be tempted to overdramatize your presentation

Language and Pronunciation. The scientific content of your presentation should be described as clearly and simply as possible, but not oversimplified. Trying to impress the audience by using sophisticated or highly technical language (something particularly prevalent among academics in the humanities!) is guaranteed to fail and is completely at odds with the aim of communicating important ideas. Thus you should try to keep your sentences fairly short (ca. 15-20 words is good) and should restrict yourself to one idea per sentence. Don't forget: the audience cannot re-hear a spoken sentence in the way that a reader can re-read the written version.

This '*one-chance-only*' aspect of oral presentations may be considered a disadvantage, but there are advantages too. For the listeners, a talk has an important '*added value*' compared to reading the same information in a paper: the speaker is better able to stress the crucial parts of the story and leave the audience with the appropriate message. Thus the information can be brought to life. But this can only be achieved if the speaker takes advantage of the opportunity to lay appropriate emphasis, as discussed above. Make full use of this possibility and don't be afraid to occasionally repeat a short statement for emphasis. You might even add a comment when you arrive at an essential point and tell the audience that this is a key part of the presentation. But do not overdo it. If every slide has such an '*essential*' point, all will be equally forgotten at the end.

If you are speaking in a language other than your mother tongue, you may need to work a bit on your pronunciation. A good understanding of the written language does not guarantee correct pronunciation. So – if in doubt – try to have a native speaker assess your pronunciation in advance. At a conference attended by one of the present authors, a French scientist described in great detail how *we eat the samples*. After several repetitions amid some chuckling from the audience, he managed to make it clear that the samples were actually being heated rather than devoured.

For non-native speakers, it is no disgrace to make small linguistic mistakes. The main concern is that you express the content clearly. Thus, also for the benefit of foreign members of the audience, do not overdo the use of long and complicated words where a simpler term would suffice. Keep to fairly short sentences. Reexpress the same idea a second time using different words. This should guarantee that you are optimally understood.

At a large international conference a distinguished professor from Oxford University began his presentation: *I apologize for not being able to give my talk in the international language of science – broken English*. Had he also made full use of the vocabulary supplied by the Oxford English Dictionary, then he will indeed have been harder to understand than many of his non-English colleagues.

To end this discussion of language considerations, let us briefly note another bad habit that is widely encountered, even among native speakers. This is the constant repetition of meaningless phrases (if you see what I mean), words *or … er … um …* noises. An *Okay?* at the end of every sentence soon becomes a distraction, as will frequent unnecessary repetitions of any kind. Your speaking should be a vehicle for communication and not a curiosity in itself! The more it resembles the latter, the less information it will convey. But such speech habits are often involuntary and the speaker may not even be aware of them. So you may need to rely on truthful colleagues to draw your attention to such weaknesses.

Running out of Time? Despite practice runs, you may find that you are running out of time (or perhaps the conference is running late and you have been asked to shorten your talk). In this case, the worst thing you can do is to significantly speed up your delivery. This will transmit less and not more information. Instead, keep to a reasonable speed and simply skip some non-essential details, ideally points that you have included for just such an eventuality. In fact, it is advisable to prepare two or more options for ending your talk, e.g. by including a small number of '*extra*' slides towards the end of the presentation. These should contain information that backs up the story already presented, but can be skipped without destroying the flow if time is in short supply. But never skip the conclusions: they are essential if people are to remember your talk.

3.5.2 Audience Contact

If you have already given some presentations, you will have noticed that everything suddenly becomes a lot easier if you manage to establish a good rapport with your audience. You feel more positive about the audience and they feel more positive about you. You have probably noticed in talks that you have attended that once a positive rapport has been established between speaker and audience, a talk automatically gets better. But like many other pyschological/ emotional entities, this '*rapport*' is very hard to define; and it is even more difficult to give explicit and foolproof instructions for generating such a positive atmosphere. However, by following the advice in the remainder of this section, you should find that your listeners – provided they are not made of stone – will react positively to your efforts to communicate with them.

Capturing Attention. Look at the audience, speak to the audience and, if necessary, react to the audience. These are the three golden rules of audience contact. Observing these rules will go a long way towards keeping the audience's attention and interest, and to establishing a positive atmosphere. Thus, to begin with, it is advisable to

Fig. 3.10. Being enthusiastic and positive will help to awaken and hold the audience's interest

stand rather than sit when giving your presentation. This way everyone in the lecture theater can see you and your voice will carry better. But even when standing, many speakers are inclined to talk with their backs to the audience or with their heads bowed over the projector. This is all wrong! Eye contact is essential. Without it you will not only lose the audience's attention but may give the impression of being insecure or arrogant. And by looking at the audience you will automatically find it easier to speak audibly rather than mumbling to yourself or to the wall.

Suppose you are attending a talk and the speaker barely manages to look at the audience. Not only do you feel that he/she is not addressing you, you also have to strain to catch what is being said. The chances are that you will soon give up the struggle, will feel under less obligation to concentrate, might start to examine the furnishings, count the attendees, look at the clock, …

If your presentation is part of your thesis defence, or is in support of a grant or project application, then you should concentrate your attention primarily on the decision-makers among the audience; it is them you need to convince. Address them directly most of the time, establish eye contact, and try to read their behavior and react appropriately.

Keeping Attention. There are two main reasons why an audience's attention may wane. Either they are genuinely sleepy or your talk is putting them to sleep. A genuinely sleepy audience is likely to be encountered in the early afternoon, shortly after lunch (or perhaps on the morning after the conference dinner – if it was a good one!) As discussed before, and shown in Fig. 3.11 a person's concentration varies in a fairly predictable manner during the course of the day.

So, if you want to have an attentive audience, try to have your talk scheduled for the morning hours, or later in the afternoon. On the other hand, if you are worried about critical questions, then these are likely to be least potent in the early afternoon.

Often, however, you will have no influence over the scheduling of your talk. So what should you do if the audience appears sleepy, yawns are being stifled and eyelids are drooping? Then wake them up! Change your speed, talk louder, open a door or window to let in some fresh air.

It may be your misfortune to inherit a bored and sleepy audience from the previous speaker. In this case, you need to show them immediately that your talk is going to be far more lively and interesting. Addressing the audience with clear pronunciation, slow speech and a readily understandable introduction is usually enough to awaken them from their stupor.

If, in the middle of your talk, you notice that some attendees, for whatever reason, have blank or even puzzled looks on their faces try to react immediately. Perhaps you have just explained something too fast. Try to adapt to the audience's needs by adding a few more words of explanation, by repeating an important message more audibly, or rephrasing the last rather complex statement.

In a short conference talk of, say, 15 minutes, there is little more you can do to ensure that the audience remains with you. But in a longer talk, lecture or seminar, some interaction with the audience can be very valuable, as can entertaining interludes. The remainder of this section considers these aspects and thus only applies to longer presentations.

In a seminar or open-ended presentation it is often useful to establish a dialog with the audience. Perhaps, when introducing the talk, you can invite the members of the audience to interrupt with

Fig. 3.11. Watch out for tired or bored attendees and react accordingly (see text)

questions and comments. An early question from the audience may well help to break the ice by generating a more informal atmosphere. When allowed to participate directly, the audience will feel more involved and be additionally motivated to follow your talk closely. Any kind of time pressure – whether it be an appointment that you must keep or a subsequent booking of the seminar room – reduces your leeway for informal discussion. Of course, even in the absence of any deadlines, there comes a time when everyone is getting fidgety and needs a break. So you must still keep enough control of the discussion to ensure that the talk does not turn into a trial of endurance for some of those present. You can always adjourn to the nearest bar or coffee machine for further discussion with the enthusiasts.

Another option for keeping the audience alert and interested is to intersperse a few anecdotes, jokes or cartoons into the otherwise serious material. A personal story is a particularly good way of capturing attention and establishing a good relationship with the audience. Tell them, for example, about the unexpected discovery that led to the whole investigation, the controversy about its interpretation, the fight with the journal referees, or how some famous colleague was involved. It will increase your listeners' interest if the work reported is given an additional personal touch that all of them can identify with.

Once I (CA) was due to give a scientific presentation at the University of Missouri. Beforehand I was introduced to several of the professors who were planning to attend. I quickly discovered that none of them was very familiar with the topic of my talk and that most of the material I had prepared would mean very little to them. To avoid the embarrassment of giving a specialist talk to a nonspecialist audience, I decided at the last minute to radically change the emphasis. Instead of discussing the scientific findings in detail, I wrapped the bare scientific bones into a story about a patent dispute between two companies. The results I had published and had intended to discuss in the talk actually turned out to be decisive in this patent dispute and had eventually led to nine out of ten patent applications being turned down. The audience listened with interest to the story about battle in court and even managed to digest some of the scientific content, which I presented as an aside. But this strategy is not to be recommended. For me it was a lucky escape. It would instead have been better to learn about the audience in advance and to prepare the talk accordingly.

Not everyone feels comfortable telling stories or jokes. So don't think that they have to be forced in at all costs. A badly told story or a joke whose punchline is not understood are worse than no diversion. As you gain experience at presenting, you will probably feel more confident about including humor. At first you may simply want to keep some up your sleeve in case of need.

If duty demands that you give a talk on a rather uninspiring topic, careful choice of supplementary material can make all the difference. A couple of years ago, at an anniversary celebration of a Max Planck Institute, the Director was scheduled to give a presentation on the Institute's 10-year history. This sounded less than promising in terms of entertainment value. However, he was a skillful presenter. By displaying interesting statistics about the Institute's publication activities, the citations of these publications, and some historical highlights he succeeded in keeping the audience captivated. In such cases, the presenter's skill and a careful preparation can turn even a dull topic into an entertaining talk.

Finally a word about lecturing to students. In this case, it is especially important that the listeners absorb and understand what you are saying. Here it may be good to ask a few questions to test whether the message has been received. If not, go over the essentials once more to fill in the gaps. You can also ask the students to pre-empt you with the answer to a problem that you have just posed. Whether or not the correct answer comes, you should then go on to state the answer clearly and to explain why it is correct. Before moving on to a new topic *'Does anyone have any problems with this section?'* offers students a valuable opportunity to clear up misunderstandings.

3.5.3 Behaviour and Body Language

In previous sections of this chapter we have dealt with two of the three main communication channels that are active during the course of a presentation. These were the visual information transmitted via your slides or other visual aids (Sects. 3.2 and 3.3) and the verbal auditory information imparted by speaking to the audience (Sect. 3.5.1). In addition to these, there is a third important channel, namely the non-verbal communication that is commonly referred to as body language. The Chambers Dictionary defines body language as *'commmunication of information by means of concious or unconscious gestures, attitudes, facial expressions, etc.'* In this section we will address those aspects of body language and behavior that significantly influence the impression generated by a presenter.

Fig. 3.12. Body language is a surprisingly important factor in determining the overall impression made by a speaker

Most presenters, scientists in particular, would rightly claim that the content of a talk should be far more important than the psychological effect of their body language. But one can find some surprising statistics to illustrate that this is not always the case.

Following a televised debate between Al Gore and George Bush in the run up to the US presidential election of 2000, a survey of members of the public tried to establish what were the main factors affecting the impression, good or bad, made by the candidates. The remarkable result was that the overall impression was determined

> ▷ to 55 % by the body language;

> ▷ to 38 % by the voice;

> ▷ to 7 % only by the content.

Whilst this is definitely not the case for a scientific presentation*, it is a clear warning: Do not underestimate the psychological effects of non-verbal communication, whether concious and unconcious.

Hence, whenever you are preparing to give a presentation, or practising one, remember that the audience's perception of you, your appearance, and you behaviour will influence their appreciation of your talk. There are several common mistakes in this respect and we will discuss each briefly.

Fig. 3.13. George Bush and Al Gore during a televised debate

Typical Mistakes. Some of the most frequent mistakes are actually quite easy to correct once you have been made aware of them. For example,

▷ Talking to the wall;

▷ Obscuring the audience's view;

▷ Being inappropriately dressed.

More difficult to rectify are ingrained and involuntary bad habits such as

▷ Displaying inappropriate facial expressions;

▷ Making nervous gestures or movements.

Let us deal with each of the above problems in turn.

At a conference you will find that a sizable portion of the presentations are given not to the audience but to the wall. The speaker is so keen to see and read the information that is projected on the wall that the audience sees only their back. Not only does this fail to keep the attention of the audience: the speaker, unless using a microphone, becomes far less audible. And a microphone only compensates partially: one should not underestimate the importance of lip-reading in assisting our understanding of the spoken word. Even those with good hearing have better speech discrimination when they can see the face of the speaker. Thus, face the audience as much as possible and read your text not from the wall but from the projector or computer monitor. If you must look at the projection on the wall behind you, then simply pause while reading it, and then turn your head back towards the audience before continuing to speak.

---

* There is an essential difference between political speeches addressed to the public and presentations by scientists for scientists: In contrast to the latter, political campaign speeches often contain many statements that listeners are not able to verify for themselves. They are simply asked to believe them. Hence the overall impression boils down to the question of the perceived sincerity of the speaker – a quality that, in turn, is very dependent on appropriate body language.

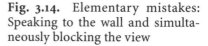

Fig. 3.14. Elementary mistakes: Speaking to the wall and simultaneously blocking the view

Beware, also, of blocking the audience's view of the projection. This can occur either by your standing in the direct line of sight between the audience and the projection (often meaning that the projection is too low), or – worse still – when you cast a shadow on the projection by standing between the projector and the wall. These are typical mistakes of inexperienced speakers, who are usually quickly asked to move. But it is better to spare yourself the interruption and distraction by avoiding the situation. When you feel blinded you should change your position.

How should you dress for your presentation? Whilst scientists are known to be fairly flexible in matters of dress, it is still better not to be totally under- or overdressed for the occasion.

Jeans and T-shirt may be your standard attire at the lab. But if you are honoured with an invitation to present a plenary lecture at a big international conference, you might consider a slightly more formal attire (shirt and tie or skirt and blouse). On the other hand if you give a talk at a company in Silicon Valley, a suit and tie would make you feel like a Martian who has just landed on Earth. When in doubt choose the middle ground of smart trousers and attractive jumper. At risk of sounding like a nagging mother: don't forget to comb your hair!

It is easy to exchange your tie for a more appropriate one, but far more difficult to change certain nervous habits. In an extreme case, such as a bad stutter, there is really not much one can do in the short term. Here probably only a long-term therapy can help. However, other nervous habits can be tackled with patience and a little help from friends/colleagues. Things to watch out for are:

Facial Expression. Avoid facial expressions that might suggest arrogance or insecurity. In fact, rather than thinking in terms of avoidance, it is probably better to concentrate on dislaying a warm smiling face which sends positive signals to the audience. Furthermore, making yourself smile also helps to induce in yourself a feeling of genuine contentment.

Nervous Movements of the Hands or Eyes. These will make the audience feel nervous too. If you can succeed in behaving calmly, this will also help you to feel calm.

Nervous Pacing up and down. Some slow relaxed movement is OK, and can even give a dynamical touch to your presentation. It is certainly better to move slowly than to stand rigidly like a soldier on guard duty. But rapid and jumpy '*dancing around*' should be avoided. Remember, you need to be aware of where you are standing; if not you will find yourself tripping over a power cable, blocking the audience's view, or disappearing from sight.

Overuse of Gesticulation. Gesticulation is hardly used at all in many Asian countries. Even where it is more common (Southern Europe), avoid making yourself the major source of entertainment, at least not if you want the audience to concentrate on the content of your talk.

Sloppy Posture. Having a hand in your pocket may make you feel '*cool*' but will not impress everyone. If you are defending a thesis or your presentation is part of a project or funding application, be sure to show due respect to your elders and all decision-makers.

Fig. 3.15. Some moderate gesticulation can help to emphasize your message. Too much will make you an unwitting source of amusement

Appropriate demeanour should not be overlooked in this respect. To help find a comfortable standing posture, you might try momentarily going up onto your toes and then returning to normal standing. This will not be noticed and will automatically leave you in a relaxed stance.

Sniffing, Head-Scratching, Nail-Biting etc. Any such nervous habit will distract the audience and give them a less than pleasant impression of your talk. Here, too, ask your friends and colleagues for honest criticism.

3.5.4 Being Relaxed

As most actors will tell you, a certain amount of stage nerves is a good thing. A slightly raised adrenalin level makes you alert and decisive. But excessive nervousness is counterproductive and the secret, both in acting and in presenting, is to get the level just right. As Vincent Di Salvo, Professor at the University of Nebraska suggests, *'Your goal is not to get rid of the butterflies in your stomach, but to convince them to fly in formation.'*

Excessive nervousness often poses one of the most difficult but important challenges for a newcomer to presenting. If you can be relatively relaxed, you will not only make a more confident impression, you will also be calmer and able to think more clearly as you deliver your presentation. Nervousness, on the other hand, may lead you to rush your presentation, to speak indistinctly, forget or confuse things.

In the above discussion of body language, we have already suggested, indirectly, that there is a definite feedback loop between feeling nervous and behaving nervously, the one causing and reinforcing the other.

This, however, also works for relaxed behavior and feeling relaxed and, with a little practice, can be used to your advantage. If you can conciously control your behavior (facial expression, movements, posture) so that you *appear* relaxed (smiling face, slow movements, relaxed posture), this will also help you to *feel* relaxed.

Being well prepared is another element which should help to avoid nervousness. If, on the other hand, you were working into the small hours the night before in order to finish your talk, overtiredness will show. Enough sleep is essential if you are to be alert and relaxed throughout the talk and the discussion.

An excellent way to relax, improve your concentration and simultaneously overcome nervous habits is to make use of *breathing techniques*. For a long time now, singers, sportspersons and yoga teachers have been benefiting from the use of simple and natural breathing techniques. The medical profession too is increasingly recognizing the contribution that good breathing can make to our general well-being. Actually this is no secret: Over the centuries and throughout the world, many religious, spiritual and other forms of meditation have become established. Yoga and transcendental meditation are two that are well known in the West. Almost all these techniques – so we claim – produce their relaxing (and even transcendental?) effect by stimulating regular, controlled and deep breathing.

By performing some simple breathing exercises now and then, you will be able to relax better, also in challenging situations such

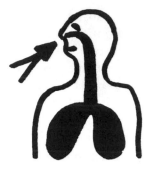

Fig. 3.16. Deep and slow breathing has a calming effect if you are nervous

as presenting a talk; overcome anxiety and sleep better; reduce chemical addictions; and improve you overall vigour and well-being.

What is good breathing? Good breathing is in fact natural breathing, for example the way a baby breathes. Essentially it is breathing into the abdomen. Such *diaphragmatic breathing*, especially when combined with mid- and upper-chest breathing is much slower and has a calming effect. It also massages the internal organs (heart, liver, intestines) helping to keep them toned and well-supplied with blood. When breathing into the abdomen, the voice becomes fuller, deeper, and sounds more agreeable. Nervous people, who breath more shallowly with the upper part of the lungs, automatically have a higher and *'thinner'* voice.

With regular practice, diaphragmatic breathing will become second nature and you will find yourself doing it most of the time at home and at work. In times of crisis (before a presentation or an examination!) an extra practice session will help you to relax and concentrate. There is plenty of literature available about breathing exercises, yoga, and other relaxation techniques. These techniques can undoubtedly promote a healthy and relaxed state of mind and body; have a look at Amazon.com for books on whichever approach appeals to you.

3.5.5 Pointing to Overheads

Returning now to the lecture theater, let us consider the possible ways of drawing your audience's attention to particular features on your slides. It certainly helps the audience to follow your argument if you point to the text or figures that you are describing orally. There are two well-established methods of pointing whilst you speak.

Many speakers and lecturers like to use a long stick to point directly at the wall projection. This is an effective method and also, we feel, one that brings a natural element of drama into the presentation. Some speakers, one of the present authors included, find that wielding a big stick in ones hand induces a certain feeling of control, both over the audience and over the proceedings in general – not a bad way to feel when you are up there having to perform. However, using such a stick for pointing requires a good awareness of your position and you must remember to turn and face the audience at regular intervals. Since you will be separated from your PC or overhead projector as you point, this method prevents you from simultaneously facing the audience and reading the text of your transparencies. So only use it when you are very familiar with

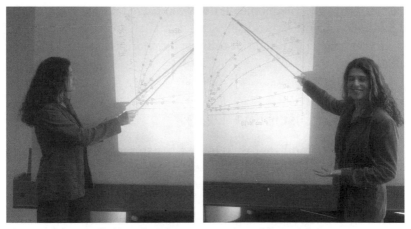

Fig. 3.17. Wrong (*left*) and right way to point with a stick

the content of your talk. Remember also to stand on the correct side of the projection, so that you don't have to point across your body. A right-hander should stand on the right side of the projection.

If the projected image is some distance away from your PC and you do not want to be constantly scuttling back and forth to change the slide, you may be able to use the same stick to cast a shadow on the projected image. To do this well for any length of time, however, requires a strong arm, as you cannot rest the stick against the wall and must not allow it to waver too much.

An alternative device for drawing attention to parts of a slide is the *laser pointer*. This allows you to accurately indicate detailed features on curves and diagrams. And, whilst you will have to turn away from the audience, you can remain standing next to the projector. But the laser pointer has its own pitfalls. If you are very nervous it will shake in your hand and be worse than useless. And if you forget to switch it off or lay it down inappropriately you might end up 'blinding' members of the audience.

There is an additional option for those using conventional transparencies. Here, you can avoid turning your back on the audience by pointing with a pen to the transparency itself, making sure that your pen casts a well-focussed and properly positioned shadow on the wall projection. You can even leave the pen lying on the transparency while you discuss the topic in question. Do not use your finger for pointing: it looks sloppy and its movement will be distracting.

3.5.6 Time Discipline

A well-planned and well-prepared talk also means a talk that can be comfortably delivered within the time available. This is especially important at conferences and other meetings with a fixed schedule. If you exceed your time limit, the subsequent presentations will be delayed and/or the breaks have to be shortened. You will not make yourself popular. It is an even greater sin to '*run over*' at meetings where there are parallel sessions. Only if the schedule is strictly kept can participants be sure of catching all the talks they plan to hear by hurrying from one lecture theater to another.

Thus as the planning and practicing stage you must be clearly aware of whether you are preparing a 15-minute or a one-hour talk. If, during your presentation, you feel that you are behind time, do not hesitate to ask the chairman how much time you have left. If you are indeed short of time, do not start to rush. It is far better to omit a part of the material you had planned to present. To cater for this eventuality, it is wise to have prepared at least one shorter alternative for finishing the talk earlier. As noted previously, this usually this means skipping parts of the results or discussion sections. But never leave out the conclusions. These represent the 'message' that you want the audience to take away.

If discussion time is scheduled, try not to encroach into this too far. A short discussion at least will help to make your talk more memorable. Occasionally it will even generate interesting new ideas.

3.5.7 Evaluation

To help you learn from your mistakes and omissions, the following checklist can be filled out by listeners hearing your practice delivery. It can also be used at a real presentation if a suitable candidate is in the audience. Finally, if you have had the opportunity to video your presentation, you can use the list yourself for self-criticism.

Fig. 3.18. Plan your timing carefully so as to avoid a last-minute rush

Checklist for Evaluating Presentations

	(poor) 1	2	3	4	5 (perfect)
Content					
Structure/organization	☐	☐	☐	☐	☐
Comprehensibility	☐	☐	☐	☐	☐
Continuity	☐	☐	☐	☐	☐
Amount of information	☐	☐	☐	☐	☐
Take-home message	☐	☐	☐	☐	☐
Slides					
Legibility	☐	☐	☐	☐	☐
Amount of information	☐	☐	☐	☐	☐
Font sizes	☐	☐	☐	☐	☐
Use of color	☐	☐	☐	☐	☐
Animation	☐	☐	☐	☐	☐
Delivery					
Overall impression	☐	☐	☐	☐	☐
Speed	☐	☐	☐	☐	☐
Audiblity	☐	☐	☐	☐	☐
Intonation	☐	☐	☐	☐	☐
Pronunciation	☐	☐	☐	☐	☐
Body language	☐	☐	☐	☐	☐
Other Comments					

3.6 Surviving the Discussion

The discussion or question session that follows most presentations will strongly influence the overall impression that you make. You may have prepared your talk very well and delivered it expertly. But you can ruin everything if you appear incompetent during the discussion. Often, especially when it comes to defending a thesis, it is the discussion that really sorts the sheep from the goats.

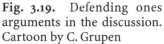

Fig. 3.19. Defending ones arguments in the discussion. Cartoon by C. Grupen

Be Prepared. There is plenty you can do in advance to be well prepared for the discussion. The most obvious measure is to really know your material inside out. Make sure you have fully understood all derivations and all logical steps in your argument. Move only on solid ground, i.e. do not present anything that you yourself are hazy about, and do not try to explain things that you do not fully understand yourself. Are there any weak links in your argument that are obvious targets for criticism? If so, think carefully about whether you really want to include this material and be prepared for awkward questions with a second line of defence. Try to anticipate other obvious questions and, better still, when practicing your talk ask your colleagues to come up with as many questions as possible. Perhaps you can even provoke certain questions by leaving a few things only partially explained in the talk. If the expected question is then asked, you might even show an additional slide which answers it perfectly, explaining that this slide was only omitted in order to save time. Both you and the questioner will make a good impression in such an exchange. Note, however, that it is a fine line between provoking questions and having members of the audience primed to ask questions. The latter is ill-advised and would reflect very badly on you if it became known that you were rigging the discussion.

Responding to Questions. It is often useful to repeat or restate a question addressed to you for the benefit of the audience. Even in a fairly small room, a question from the front row may not be heard at the back. In large lecture halls such repetition is essential, even though microphones may be provided for audience questions. By repeating a question you also buy yourself some extra time to compose your answer. Furthermore, if the question is poorly expressed or has a negative/critical slant you can rephrase it slightly for clarity and to allow you to give a positive response.

Many, indeed most, questions will simply demand some clarification or further information about the work that you have presented and, if you are well prepared, it should be possible to give a concise and clear answer. But some types of question can pose more of a challenge: Generally these will be complex questions that touch on a field you know little or nothing about. Ideally you will still know the answer because you will have at least scanned the relevant literature. But if you don't, a legitimate course of action is to refer the question to one of your co-authors in the audience who is likely to know more. Alternatively, you can simply give some further elucidation of the relevant aspect of your talk, hoping that this will illuminate the issue and that the questioner will not be so persistent as to fault your answer or repeat the question.

In general, if it is really beyond your knowledge and ability to answer a particular question, it is better to be honest and simply say *'that is an interesting issue, but I'm afraid I can't give you a good answer now'*. Honesty will be appreciated more than insubstantial waffle. If the question pertains to a field that is outside your expertise, you can sometimes turn the question back to the person who raised it with a remark such as, *'that is beyond my competence, but perhaps you can provide some enlightenment'*.

If you understand the question but are unable to quickly formulate a convincing answer, the best course of action is to say: *'That is a good but rather complex question and would take a long time to answer. Perhaps we can discuss it in the break'*. This tactic can also be adopted when you know that you should be able to answer a question but have not previously thought it through or even when you have a blackout due to nerves.

Fig. 3.20. Interruptions, if dealt with quickly and calmly, should not disturb the speaker

Also challenging are questions that are quite obviously nonsense, those that are incomprehensible, and those that are repeatedly inaudible. Here, rather than publicly criticising the questioner for his stupidity, poor English or timidity, it is sometimes better to adopt the policy mentioned above for difficult questions. Simply explain a little more about the relevant aspect of your work and trust that this will satisfy. If not, other members of the audience, or the chairperson, may well come to your rescue.

Bear in mind that questions are often asked not because those asking are really interested in hearing a detailed answer. As often as not, they are asked so that the questioner can register his presence and impress upon the rest of the audience that he too knows a lot about this field and has some good ideas. In this case the answer is of secondary importance and should simply sound confident and reasonable. As we all know, when politicians are asked a direct question they rarely answer it directly, preferring to make some general nonspecific statement or to answer a completely different question. This '*skill*' can be useful to scientists too, although it should only be employed in an emergency.

Fortunately, most questions asked are sensible and fairly harmless. Provided you understand the research you are describing, you should have no trouble in answering them adequately.

Interrupting Questions. Don't worry if a brief question comes from the audience in the middle of your presentation. Usually it is just a matter of giving some brief clarification, whereafter you can continue immediately with the planned material. But don't get trapped into a lengthy discussion by such an interjection or you will run out of time and fail to present essential parts of your talk. Such questions should be countered with: '*I would like to spend more time on that question at the end.*'

3.7 The Art of Asking Questions

Most of the questions that follow a presentation are quite genuine and stem from an interest in the scientific ideas discussed. If such a question occurs to you when you are in the audience, don't be too shy to ask it: Its answer may really help you with your research and the point you have thought of may be a very good one. Bad questions are quickly forgotten; all that will be remembered the next day is that you took part in the discussion. And as the saying goes, any publicity is good publicity.

A minority of scientists – but usually one or two in every conference session – take their publicity very seriously. They see the discussion session is seen as a chance to perform. For a brief moment the questioner becomes the focus of attention and has the opportunity to show that he knows a lot and has good ideas. Some years ago, in an evening session of a meeting of the German Physical Society, there was an amusing presentation by Josef Lieberetz (Bonn) on '*Scientific Dialogology*' which classified the suspect methods of asking questions. So, for those wishing to ask questions at all costs, here is a cynic's guide to ways of devising questions (also useful to session chairmen under some circumstances):

The Examining Question. When an attendee believes that he knows far more about the topic of a talk than the presenter himself, he tries to prove this by catching the speaker out with an examining question.

Method of Modified Boundary Conditions. This method can be applied even by attendees who understood nothing of the presentation. For example, if the experiments presented were performed at 500 K and no explanation given, one might ask '*What would happen at 600 or 400 K?*'

Fig. 3.21. Trying to appear important – a not entirely laudable, but relatively common motive for asking questions

Intentional Misunderstanding. This method reflects less well on the questioner and should only be used in an emergency. If the speaker said that a law applies at normal pressure, the question might be formulated as: '*If I understood you correctly, you say that this behaviour is only observed above normal pressure. But our measurements show similar effects at normal pressure too.*'

Method of Autapotheose. This means simply '*self-praise*'. It is encountered surprisingly often and can be identified in questions beginning with a reference to the questioner's most recent paper in Nature, his plenary talk at a (much larger) international conference, or his conversation with his friend, a named Nobel Prize winner. Its purpose is to make clear that his standing is immeasurably higher than that of others present.

The Sceptical Question. A preferred method of older scientists who are envious of the success of younger colleagues. Usually the question attempts to cast doubt on the validity of the results or conclusions. It only works when posed by a senior scientist to someone more junior. When a leader tells a newcomer that he cannot understand the explanation just given, this means that the explanation was poor. Reverse the roles and the questioner – now the junior – simply casts himself in a bad light for not being able to follow the expert.

Method of Deviation. Here the questioner leads the discussion to an adjacent topic that was not actually important in the presentation, but about which he is more expert and can overwhelm the speaker with his profound specialist knowledge.

Prepared Questions. For newcomers wishing to ask questions, it is possible, with the help of the abstract to invent questions in advance. A very unusual question of this kind arose once at a conference following a presentation by the president of the American Physical Society. A member of the audience stood up and said '*I have one question.*' He then proceeded to read our a four-page manuscript and finished with '*Please explain to me why this excellent paper was not accepted for publication in Physics Review Letters!*'

The Stupid Question. Especially dangerous for the speaker are questions that begin '*This is probably a stupid question, but ...*'. They are typically asked by experienced professors who are not specialists in the field, but have the ability to view the work described within a much broader context. In contrast to most of the other types of question listed, such basic and penetrating questions are usually anything but stupid. And if the speaker has the misfortune not to have thought about the matter, he may find himself in an embarrassing situation. (Often the questioner himself will come to the rescue.)

3.8 Poster Presentations

3.8.1 A Curse or a Blessing?

Having submitted your promising abstract to the conference organizers, you may be disappointed to learn that you have been allocated a poster rather than an oral presentation. Others are inclined to be relieved because it is somewhat less stressful to give a poster contribution. Although posters have often been considered

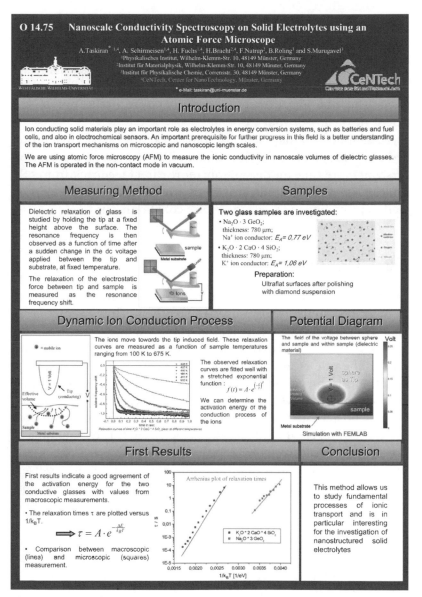

Fig. 3.22. A good poster

less important than talks, this is not necessarily the case. It is true that they tend to be assigned to more specialist contributions, but they can still report on the initial steps of some really ground-breaking research. Many conferences simply receive so many abstracts that only a fraction of these can be assigned oral presentations. Thus the allocation of a poster is rarely a reflection on the scientific quality of your work.

Remember too that there are some clear advantages to a poster session: Apart from being less demanding to present, a poster offers you more time for discussion, more direct feedback from other scientists and can thus give you more ideas for future work as well as providing the seeds for future collaborations.

3.8.2 Optical Features

For a poster presentation to be fruitful, your poster has to be noticed among, perhaps, a hundred or so others. This, initially, is not so much a question of content as of optical impact.

The first, and frequently only, part of a poster that a visitor will read is the title. And even this won't be read if it is too long, too small, too technical or just plain illegible. Thus it is essential to choose a concise title and to write it in large clear letters at the top of your poster (sometimes title banners are provided, but it does no harm to repeat the title on the poster itself). The remainder of the poster should contain as little text as possible. Concentrate more on graphics and images, keeping these as simple and self-explanatory as possible. Only in this way will a passer-by have the patience to actually read your poster to the end. Since the text should be very short (max. 1 page in total), it should include only the most important aspects of the work – predominantly, therefore, the conclusions. Keep any description of experimental and theoretical procedures to an absolute minimum. Results, too, should be presented as concisely as possible. If your results lead to interesting conclusions, other attendees will be motivated to ask you for more details.

When designing the poster, pay particular attention to good quality graphics. A4 size figures are to be recommended. If possible, the whole poster, which will generally be about size A0 (84 cm x 119 cm), should be printed on a single sheet. Not every institute or university will have a printer that can do this, but a copy shop can usually oblige.

Some conferences offer a prize for the best poster. Here scientific content, style of presentation and optical impact all play a role. Where such a prize is awarded, aim to win it! In addition to a possible bottle of champagne or $ 100, it will encourage you to think carefully about designing your poster.

3.9 Some Tips for Chairing Meetings

3.9.1 Conference Sessions

It is quite an honour to be asked to chair a conference session. If nothing else, it means that you have established a good reputation in the field concerned. But usually it also indicates that you are held to have good communication and diplomatic skills. So as not to disappoint this trust, there are a few rules that should be heeded.

Preparation. Before the session begins, make sure you are familiar with the schedule and the intended lengths of the presentations. Can you pronounce all the names of the speakers? At the start of the session a few words introducing the topic of the session and setting the scene for the talks that follow will help to focus the audience's attention and put you firmly in control of the proceedings.

Time Discipline. Thereafter, one of your most important tasks as chairperson is to take care of time discipline, ensuring that each speaker begins and ends at the scheduled time. This is especially important in a conference which is holding parallel sessions. You may like to use a timer which sounds a warning say 2 minutes before the scheduled end of the presentation. If the speaker fails to finish in this time, you must decide whether (a) to let the talk run on into the discussion time or (b) interrupt and ask the speaker to close immediately. Which of these options is better is a matter for your personal judgement (How interesting is the talk? How important is the speaker? Are their likely to be many questions?). If the session is unavoidably delayed, e.g. due to technical problems, you should try to get it back on schedule by asking speakers to shorten their talks slightly and/or reducing the time for discussion.

Guiding the Discussion. The other essential responsibility of the chairperson is to guide the discussion after the talk. Once the audience has been invited to give their questions and comments, you should endeavour to select people in the order in which they register their wish to contribute. You should try to allow every question to be answered to the satisfaction of both questioner and speaker before moving on to the next one. If the audience initially fails to come up with any questions or comments, it is common practice for the chairperson to ask the first question. Sometimes this stimulates further discussion, but not always. In any event, as chairperson, it is advisable to think up one or two sensible questions while the talk is in progress. In the event that there are many questions and time is running short, the chairperson should give priority to people who have not yet contributed. Eventually, she or he may have to close the discussion and suggest that it be resumed in the next coffee break.

Cancelled Talks. What should you do if a talk is cancelled or a speaker fails to show up? Provided there are no parallel sessions and the next speaker is available, you can simply proceed with the next talk ahead of time. If, however, no speaker is available or you believe that many participants of parallel sessions will miss talks they wanted to hear, you should keep to the original schedule and simply allow and impromtu break and/or a continuation of the previous discussion. If you know of the cancellation well in advance, you can even give each speaker a little longer for their talk, e.g. 20 instead of 15 minutes.

3.9.2 Seminars and Internal Meetings

At less formal internal meetings and seminars the organizer and chairperson are usually one and the same person. Here it is important to think about matters such as planning the meeting well in advance, avoiding conflicts with other events, announcing the speaker(s) and topic(s) in good time, reserving the room, etc. When the meeting actually takes place, the atmosphere is generally fairly relaxed. There is not such a need to maintain time discipline, but the length of the presentation should be agreed in advance with the speaker. A typical length for an internal meeting is 90 minutes including the discussion. The chairperson must have the courage, when necessary, to curtail lengthy and less relevant discussions. If the discussion looks as though it will continue into the night, it is often good to officially declare the meeting closed so that those who want to leave can do so. This still allows those interested to remain, or adjourn to the nearest bar.

3.10 Dos and Don'ts

DO: Know your audience.
DON'T: Give a specialist presentation to a lay audience.

DO: Select an appropriate amount of material for the time available.
DON'T: Try to squeeze in all the details.

DO: Plan at least 2 minutes for each slide.
DON'T: Include images or text that you won't have time to discuss.

DO: Structure your talk logically (similar to scientific paper).
DON'T: Jump from topic to topic.

DO: Keep to one topic per slide.
DON'T: Overload the slides with information.

DO: Use keywords as a skeleton on the slides.
DON'T: Write complete sentences or long chunks of text.

DO: Make attractive slides using some colour.
DON'T: Use rainbow coloured or black and white slides.

DO: Acknowledge those who have supported the work presented.
DON'T: Indulge in plagiarism.

DO: Draw clear conclusions; provide a *take-home message*.
DON'T: Conclude only with speculations.

DO: Practice your presentation with colleagues, friends and relations, and/or alone.
DON'T: Assume *It'll be alright on the night*.

DO: Listen to the criticisms of colleagues.
DON'T: Repeat the same mistakes time and again.

DO: Speak naturally and freely.
DON'T: Read your talk from a script.

DO: Speak and pronounce words clearly. Practice the English if necessary.
DON'T: Speak too fast, too quietly or without attention to good pronunciation and grammar.

DO: Pay appropriate attention to your dress and demeanour.
DON'T: Block the view to the overhead projection.

DO: Try to appear (and feel!) confident and relaxed.
DON'T: Appear arrogant or indulge in irritating nervous habits.

DO: Finish your presentation within the scheduled time.
DON'T: Ignore the chairman's request to conclude.

DO: Prepare in advance for the discussion
DON'T: Answer a question until you are sure that the whole audience has heard it.

We hope that these hints concerning effective scientific presenting will be helpful. In particular, we trust that the pointers to the most frequent mistakes of presenters will help you to learn the smart way:

The stupid guy never learns.

The average guy learns from his mistakes.

The smart guy learns from the mistakes of others.

4 The Culture and Ethics of Scientific Publishing

Before turning, in Chap. 5, the matter of writing a good paper yourself, here we give a more general exposition of scientific publishing, its purposes, the forms it takes and some associated ethical questions. Those readers mainly interested in advice about preparing papers, can – at risk of missing out on some interesting discussion – proceed directly to Chap. 5. Details about electronic publishing can be found in Chap. 6.

4.1 Purposes of Scientific Publishing

In this section, we will show that scientific publishing serves many different purposes. These can be summarised as follows:

> Documentation of knowledge

> Discussion forum

> Motor for further progress

> Seeking and retrieving information

> Teaching

> Establishing priority

> Furthering a scientist's career.

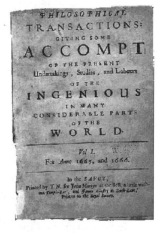

Fig. 4.1. Documentation of scientific results

4.1.1 Duty to Publish

The task of a scientist is to generate/discover new knowledge. Most scientists enter their field – be it theoretical physics or molecular biology – because it awakens their curiosity and they experience satisfaction in understanding and advancing our knowledge in their chosen domain. However, modern scientific research does not take place and receive funding simply to provide intellectual pleasure to the scientists involved. Its overall aim is to contribute to the commonly held pool of scientific knowledge, with which we aim to make the world a better place.

Thus, if scientists were to perform research without publishing their results, it would be like a company manufacturing a product but not selling it. And like the company, which will go out of business if it does not sell its product, a scientist who does not publish his or her research will also quickly lose the support needed to continue working. And since most funding comes, directly or indirectly, from public taxation, scientists also have a moral duty both to produce and to distribute their wares.

Through publications in scientific journals, new knowledge is made available and recorded for the use of present and future generations. Thus every publication contributes a small building block to the edifice that is Man's knowledge of the world.

4.1.2 A Motor for Scientific Progress

Scientific publications serve as an important forum for information exchange and discussion. Comments and further papers by other authors are often necessary before a consensus can be achieved about the interpretation of new results. Another good reason for publicizing ongoing research is to avoid duplication of effort: reading the literature before beginning a research project will help you to avoid reinventing the wheel.

Without appropriate publications, many important new fields of endeavour would never gather the momentum needed to drive sustained progress. One need only consider the discovery of x-rays

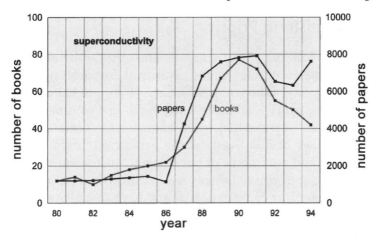

Fig. 4.2. Annual numbers of books and articles published in the field of high-temperature superconductivity

or of high-temperature superconductivity: seminal papers in these fields started an avalanche of activity. The effect of the 1986 Bednorz-Müller paper on high temperature superconductivity[*] can be seen clearly in Fig. 4.2.

4.1.3 Teaching and Reference

Scientific publications, especially in book form, also play an important role in teaching at all levels. From a school pupil's first encounter with a science lab through to the most advanced graduate courses and the self-teaching of experienced scientists – the clarity and quality of presentation of the accompanying written information (school books, university text books, scientific monographs) can strongly influence the success and enjoyment of the learning process.

Reference books also play a special role in science. This is because of the sheer amount of data and other information that science has generated and which is needed – often on a daily basis – by other scientists in their work (see also Sect. 4.2.4).

---

[*] J.G. Bednorz and K.A. Müller, Z. Phys. B **64**, 189 (1986)

4.1.4 Establishing Priority

Another important function of scientific, and other academic publishing is as a means of establishing priority and of protecting intellectual property. When a new idea is first published, the author(s) of the paper establish themselves as the originators of this idea. It should be noted that, when it comes to determining priority, it is the date of submission of the paper reporting the idea that is decisive and not its date of publication. Most scientists care a great deal about priority of discovery. But this is not surprising. Being credited with an important advance can have an enormous influence on your career in science. And of course it is also a matter of prestige. New phenomena, new processes, or indeed new diseases are frequently named after their discoverer, i.e. the person who first reported them in the literature. Think of the Planck constant, the Haber process, or Hodgkin's lymphoma, to name but a few.

Fig. 4.3. Nobel Prize diploma

The Nobel Prize committee also requires evidence of priority, in the form of published work, and sometimes one can connect a major publication to a subsequent Nobel Prize.

During the year 2003 a tough competition developed to be the first to produce a working atomic laser, perhaps an achievement that will be honored with a future Nobel Prize. Two groups were working on this, one at a Max Planck Institute in Munich and the other at the National Institute of Standards in Boulder, Colorado. For some while they progressed neck and neck and finally both were ready to publish their findings. As it happened the two papers were submitted just one day apart. It will be interesting to see whether this has a significant effect on the future distribution of laurels.

In centuries gone by, there was a trick that scientists could use to 'reserve' priority, even if they were not yet ready to publish their ideas in a comprehensive and confirmed form. Authors were allowed to send their manuscripts to a journal as a so-called *pli cacheté* or sealed envelope. The date of submission, essential for priority purposes, was the date that the paper arrived with the publisher. However, the author reserved the right to tell the publisher when the envelope could be opened and the commnication published. This was a simple and effective way of laying claim to an idea without having to publicize it immediately; useful, for exam-

Fig. 4.4. In days of old, a sealed and dated envelope could be used to claim priority

ple, when the idea was still awaiting some further experimental corroboration, or when one wanted to keep competitors off the track.

This practice was used by many members of the *Académie Royale des Sciences* in Paris in the 18th century. In 1747, for example, Alexis-Claude Clairaut deposited four plis cachetés and Jean d'Alembert deposited two of his own as they (and Leonhard Euler in Berlin) had all been working simultaneously on the three-body problem, particularly as it applied to the motions of the moon, Jupiter, and Saturn.

But it turned out that this system was flawed. It was misused by one author who came up with two plausible but different solutions to a certain problem. Not knowing which to prefer, he submitted both as plis cachetés. The trick was revealed when, later, having established the correct solution, he mistakenly told the publisher to open the wrong paper. Recognizing his mistake, he tried to change this decision, but thereby drew attention to the swindle, eventually precipitating the end of the *pli cacheté*.

4.1.5 Other Benefits to the Individual

All active researchers rely on the published literature to keep them informed about what is going on in their field. Students beginning their postgraduate studies and scientists entering new fields also need to survey the literature to establish the state of the art in their chosen subject.

Without careful checks of the literature, it is likely that many PhD theses and papers submitted for publication would more or less duplicate previous work; a fact that would hopefully be recognized when they are evaluated by referees. To avoid wasting time reinventing the wheel it is essential to regularly scan the relevant publications and, if available, make use of an electronic alerting service to tell you when a new paper on your subject is or will be published.

The availability of our accumulated knowledge and information in written form, whether printed on paper or electronic, is also a prerequisite for scientific and other intellectual progress. In the days before printing, most knowledge was passed down from generation to generation by word of mouth.

Powerful though our brains are, their capacity to store and recall information is still small in comparison with the modern availability of written information. By committing our knowledge to paper – or other means of storage, we free up our brains for more creative activities than simple memorising. This is an important reason why the pace of scientific progress has increased so much in recent times.

Last but not least, the publishing of ones ideas and findings is essential for every scientist interested in establishing a good reputation and advancing his or her career. When you have reached the stage of receiving frequent invitations to give plenary lectures at big international conferences, to write review articles and book chapters, to chair conference sessions, then you may consider yourself well established. Until such time, it is wise to pay careful attention to your publishing activity. The well-worn phrase 'publish or perish' contains a good deal of truth. Like it or not, an applicant's publication and presentation list is often one of the main criteria for selecting candidates for academic positions. It is also carefully scrutinised when scientists apply for funding. The better your publication list at a particular stage of your career, the stronger your chances of progressing to next level.

There is no firm rule about how many publications are to be expected when, but some rough figures can be given. Please be aware that these figures may also vary from subject to subject and from country to country.

Many students gain their master's degree without authoring any original publication. But a good master's thesis often is – or could be – the source of at least one publication, co-authored usually by the student and his advisor.

A doctor's degree, based on a more extensive research project, is usually accompanied by one or more original papers reporting on major chunks of the work, perhaps two conference contributions (proceedings papers based on talks or poster presentations), and sometimes some further short notes in periodicals.

Fig. 4.5. Science Citation Index – measures the impact of publications

As a postdoctoral assistant (or assistant professor in the US) on the way to tenure, you will continue to gather publications. If you eventually aim to gain a full professorship, then typically your publications list should contain at least 20 full-length original papers, 50 short notes and one review paper. It is also advantageous to have authored or contributed to a scientific monograph or edited volume. In this way you have documented the milestones that you reached in the course of your research.

However, you should not be misled into thinking that only the quantity of publications is important. An additional criterion that is increasingly used nowadays to evaluate candidates' publication lists is to check the impact of these papers. This latter can be established with the help of the science citations index, which records, among other things, how many times a particular paper has been cited in other publications. Highly cited papers are naturally more highly regarded and count more in a candidate's favor than a paper that has never been cited.

For a professor with a permanent position the pressure to publish is released a little. This is a good thing too, since a professor must generally also find sufficient time for teaching, fundraising, and managing a research group. Fortunately he or she then begins to profit from the fact that his students and assistants are actively researching. His job, as sounding board, mentor, provider of ideas, etc., qualifies him to become co-author on many of the papers written by his students and junior colleagues. If successful at this job, a scientist can fairly claim to have '*made it*' in the sense of achieving a sound professional standing in academic and scientific circles.

4.2 Types of Publications

As a preliminary to giving details about publishing your own work, it will be useful to briefly survey the various types of publication that are available. We will begin with the simplest and fastest, but least established, and move on to more conventional, better recognized, but more demanding.

4.2.1 Report on a Homepage or Preprint Server

The easiest and fastest way of making your discoveries available to the public (which is the true meaning of the term 'publish'), is to store them on your own, or your institute's, homepage.

Making your results available in this way has the disadvantage that only a small circle of people are likely to stumble upon them. Furthermore, without peer review, the content of such papers may not be given the same credence as work published in peer-reviewed journals. Thus, whilst homepage reports are a useful additional way of drawing attention to yourself and your research, such *pseudo-publications* are not to be recommended as the only means of communicating your work, especially if you expect it to be of interest to a wide audience.

On a well-organized and recognized preprint server there is a significantly higher probability that people will find and indeed cite your report. Currently the best-known preprint server for physics is the e-print archive http://arxiv.org, which was formerly maintained by Los Alamos National Laboratories, but is now with Cornell University. It is as well-organized as a scientific journal and regularly frequented by many researchers. A disadvantage of this and most other preprint databases is that the material is still not peer-reviewed.

Thus the results have not necessarily been scrutinized and approved by independent experts in the way that they are in journal publications. Only the papers that later appear in recognized peer-reviewed journal can be considered to be of approved quality. Such reports are also not citable in the same way as publication in a book or journal; and alone they do not qualify as an entry on your publications list. Priority of discovery is another problem that may arise if the relevant report appears only on a preprint server.

4.2.2 Conference Proceedings

The material presented in the talks, and sometimes on the posters, at a conference is frequently written up and documented for posterity in the form of conference proceedings. These proceedings may be published as a regular or special issue of an appropriate journal, as a book issued by a recognized publisher, or simply as a bound set of manuscripts, copied and distributed by the institute hosting the conference.

Reporting your results in a conference proceedings that is published in a recognized journal or a book with an ISBN is the easiest way to gain a citable* scientific publication. Some conferences, for example the Japanese Conference of Applied Physics prepare only an abstract booklet for distribution to the conference participants; an abstract in such a booklet is not a citable publication.

The papers accepted for a conference proceedings are often strictly limited in length, with typically 8 pages for a plenary talk, 6 pages for an invited talk, and 4 pages for other contributions. Due to the limited space and relatively low circulation of many conference proceedings, it is a common and accepted practice for the same material to be presented in more detail later in a longer journal paper.

---

* The condition for citability is that the paper appears in a book or journal bearing ISBN or ISSN number, respectively. ISSN means International Standard Serial Number and is the identification code for periodicals. Every scientific journal or book series receives an eight-digit ISSN number from the International Serials Data System in Paris. It should be noted that ISSN numbers change whenever there is even a slight change in the title of a periodical.

For books, the corresponding ISBN (International Standard Book Number) is issued by the Library of Congress in Washington. This ten-digit number identifies the language domain of the publisher, the publisher itself, and the individual book, ending with a check digit to help avoid errors.

4.2.3 Journal Publications

For the rapid dissemination of new knowledge, its publication in the form of a journal article is almost certainly the most important and effective channel.

The acceptance of a paper in an established peer-reviewed journal also provides a kind of 'quality control'. For this purpose the journal – or rather its publisher or scientific editor – subjects each paper to a process of refereeing by one or more experts, whose approval is necessary before the article can be published.

These days, many journals are available both in conventional paper form and also electronically. Some few journals are electronic only, an idea that is still at a rather experimental stage.

4.2.3.1 Types of Journal Article

The papers published in conventional scientific journals can be broadly classified into three main types:

Review papers are relatively long and comprehensive articles reporting on the state of the art in a particular field. They are usually authored by invitation by leading experts in the field. Frequently, one or two review papers will be published as the main attraction in each issue of a journal. Some journals are dedicated to the publication of review articles, for example Review of Modern Physics. A comprehensive review is usually anything between 20 and 100 pages long, i.e. it may take on the nature of a short book. Review papers do not usually report on the author's latest discoveries, although his or her previously published work in the field may often feature prominently. A good review paper should endeavour to cover the field as comprehensively as possible, beginning with the pioneering early work in the field and attempting to cite all recent work and publications. For newcomers to a field a review article can be extremely valuable, especially if it carefully evaluates the various contributions to a topic. This is an enormous help, especially to students. The scope and purpose of a review paper is thus quite similar to that of a book. It differs mainly in that the discussion of basic concepts and background knowledge cannot – for reasons of space – be as detailed as in a book. Because of their wide-ranging content and appeal to a broad readership, review papers are, on average, cited much more frequently than other journal articles. One reason for this is that lazy authors often choose to cite a review paper in place of the large number of original publications on which it is based.

Original papers are the main channel of communication for the active scientist. They present significant new results together with a detailed description of how they were obtained (experimentally or theoretically) and an analysis of their implications. Such a paper is usually prepared when a research project has reached a fairly advanced stage of completeness and the author is able to provide all the details necessary to understand what can be a fairly complex new set of facts and relationships. In addition to the presentation of the author's own results and their interpretation, it is important that adequate comparison and discussion is given concerning related, and even conflicting, results of other researchers. Likewise adequate references to related publications should be included. The length of such a paper can vary a lot, but as rough guidance, 10 pages is often appropriate.

Letters or short notes are often the publication of choice when one wants to briefly and rapidly communicate important new results and establish priority. They also often serve to awaken interest in a new field of investigation. Frequently a letter is published before the full implications of the research are known. After some further follow-up work, a more detailed report including additional results and deeper discussion is often published as a full-length original paper. A letter or short note is frequently restricted in length: depending on the format of the journal concerned, a typical length is one to five pages.

These days, journals are not only available as printed paper documents: most established international science journals (especially in the physical sciences) are additionally available in an electronic version, which can be accessed via the internet. The topic of electronic publishing will be discussed in detail in Chap. 6.

4.2.3.2 Leading Journals

First we should consider the question: What qualifies a journal as 'leading'? Well, the main available criterion is its ranking in the Science Citations Index, published by the Institute of Scientific Information (ISI)*. In this index, journals are grouped by subject and evaluated according to the average number of times that the papers in each journal are cited in other publications. This yields the impact factor for each journal; the higher the impact factor, the higher the journal ranks in its category. The exact definition of the impact factor of a journal is the average number of citations per article within the second and third year after publication of the article. As such, the impact factor clearly reflects both the number of people reading the articles and the significance of the information presented in them. But, as with all statistics, one should be wary of overinterpreting the impact factor. For example, it turns out that impact factors are almost completely determined by the 10% of most cited papers. Even in high-impact-factor journals, the majority of articles are rarely cited. Sometimes an article dealing with a special topic will reach its intended readership better if it is published in a lower-impact specialist journal than in a more general higher-impact journal.

4.2.3.3 General Science Journals

The best known general science journals are '*Nature*' and '*Science*'. Only the most exciting results that are of interest beyond the confines of a particular discipline are considered for publication. Newcomers are not recommended to submit there first papers to one of the top journals – unless, that is, your professor began to jump up

---

* The ISI also publishes many other statistics and derived information. For example, one can find the average half-life for each journal and lists of highly cited authors. However, a full description of this data and its potential uses and abuses would take us too far afield.

Fig. 4.6. *Nature* and *Science* – arguably the most prestigeous scientific journals

and down excitedly when you first showed him your results and is now joining you on night shifts to complete the data-taking.

As Roger Highfield, science editor of the London Daily Telegraph put it:

> *'Most scientists would murder their granny to get into Nature or Science. These are very high-profile, high-prestige scientific journals.'*

There are numerous standard journals in the various main disciplines with a high-impact factor. Review journals in particular tend to cover a fairly wide area and are frequently cited.

4.2.4 Books

> *'Don't know much about a science book …'*
>
> Art Garfunkel, lyrics to *'What a Wonderful World'*

Books represent the most detailed and comprehensive form of scientific publication. In a book, the knowledge previously distributed over various journals and hundreds of articles can be brought together within a common framework, providing detailed background and yielding a comprehensive picture.

Whereas a journal publication usually focuses on the presentation of particular progress and assumes that the reader is already familiar with the state of the art, a book can (and generally should) also give a tutorial introduction, providing access to the topic for non-specialists too. It can be more detailed in all respects and will, ideally, leave the reader understanding how the topic concerned fits into the greater scheme of things. In general, books fulfill quite a different role to journal publications. In a journal, one finds smaller pieces of 'fluid' information, i.e. progress reports in fields that are evolving. A book, on the other hand, usually presents more established or 'solid' knowledge which has become widely accepted. Furthermore, a book can and should explain the basics at a level accessible to all targeted readers, and also give a good deal more background information on the topic at hand. It is longer, more detailed, and more comprehensive.

As with journal publications, one can distinguish several different genres of scientific books:

Monographs are written in a didactic and uniform style, by one or a small number of authors, each an expert in the field. The book is written jointly – i.e., all chapters are the responsibility of all authors. If the writing is divided among the authors, the text is later revised for consistency and completeness so that it is impossible to see the joins. Such books are designed to document milestones in a particular field of research. They are usually aimed at a readership of researchers and advanced students seeking access to a new field or simply a deeper and more systematic understanding in their own field.

Textbooks are similar in scope to monographs, but with greater emphasis on the tutorial component. Assuming a fixed level of background knowledge (e.g. a bachelors degree in chemistry) they continue from this point to provide all the necessary tools and information to give the reader a clear understanding of a particular branch of science. Good didactic skills are usually needed to write a good textbook. The value of a textbook is also much enhanced if it contains exercises on all the key topics. Whereas students like to have exercises with answers or at least hints for solution, teachers and lecturers are usually more inclined to recommend books in which the answers are provided separately and only to teachers.

Multi-author books, almost always co-ordinated by one or more volume editors, are designed to give a state of the art report on a fairly broad field of investigation. Each chapter is contributed by one or more experts and, in itself, has very much the character of the articles in review journals. A careful editor plans the book in advance and co-ordinates the work of the chapter authors so that the whole book gives a consistent and comprehensive view of the field in question. Such a book can be just as coherent as a monograph. Frequently, however, such books are not as consistent (e.g. in notation and terminology) as they could be, and are thus recommended mainly as a source of further information for libraries and established researchers.

Reference books and handbooks are intended to present, in a well-structured form, all available knowledge, data and techniques relevant to a particular area of endeavour. The former tend to be weightier and more expensive, but usually cover a fairly wide field (semiconductor physics, say) and are usually found only in libraries. Handbooks may cover a narrower domain (e.g. Auger spectroscopy), but are designed for daily use by the practicing scientist.

Proceedings books need not be further discussed here, since they have already been mentioned above. They contain short journal-like papers bound together in book form.

4.3 A Few Words About Ethics

What type of material is suitable for publication as a scientific paper? What constitutes an ethically dubious publication? To answer these questions and establish some general guidelines, various national and international societies, including the American Physical Society and the German Physical Society, IEEE, the American Chemical Society, the American Mathematical Society, the National Science and Technology Council, and the White House Office of Science and Technology Policy (OSTP), have developed and agreed upon a *'code of honour'* [APS Ethics and Value Statements, 2002].

This code tries to define a culture of scientific publishing and to ensure that honesty is a prime characteristic of the enterprise.

The American Physical Society Guidelines for Professional Conduct

The Constitution of The American Physical Society states that the objective of the Society shall be the advancement and diffusion of the knowledge of physics. It is the purpose of this statement to advance that objective by presenting ethical guidelines for Society members.

The following are minimal standards of ethical behavior relating to several critical aspects of the physics profession:

Research Results

Publication and Authorship Practices

Peer Review

Conflicts of Interest

Co-authorship

Other Statements

http://www.aps.org/statements/02.2.html

http://www.aps.org/statements/02.3.html

Fig. 4.7. Outline of, and topics covered by APS current ethical statements

In brief, the code can be paraphrased as follows:

> *Publish substantial and new results only. Publish original results only, avoiding any republication of results already published elsewhere.*

The content of your papers is more important than their number. Unfortunately there is a tendency to aim for quantity rather than quality, one reason being that external funding agencies use publications as a measure of the success of a research project. This often tempts researchers to republish more or less the same results in several publications, each time only slightly modifying the presentation and discussion. This contributes to an unnecessary flood of publications, which would be better avoided. Who is going to read these rehashed versions of the same material? It is likely that the same people will read all of them, and will be disappointed by the lack of substantial new information.

In physics alone, there are about 250,000 new papers published each year. Supposing that only 1 % of these are relevant to your research field, you would still have to read 50 papers each week to keep up. By avoiding unnecessary publications, you thus help to make the work of other scientists more efficient too.

> *Publish true and confirmed results only.*
> *Do not falsify or invent data.*

Whilst a novelist or poet is free to invent the circumstances that he or she describes in writing, scientific writing is assumed by its readers to be factual, i.e. to truthfully describe observations and accurately report on calculations. Nor should it omit uncomfortable facts, known from other work, that contradict the conclusions being drawn.

Thus accuracy is more important than speed. Theoretical results should be carefully checked before they are published, and experiments repeated to check the reproducibility of the results. Extraordinary results – should you be fortunate enough to acquire any – demand extraordinary supporting evidence. The burden of proof for a truly unexpected or major new discovery is therefore much greater than for a more routine measurement that simply extends the domain to which an existing theory applies.

If, for whatever reason, you nonetheless publish a paper that contains a significant error, in the data or in the interpretation, it is advisable to submit an erratum. Your future work will probably build on the present foundations and so it is important to set the record straight as soon as possible. If you continue to publish papers including the same mistake, you are in danger of losing credibility in the long run.

▷ *Accept the concepts of intellectual property and copyright and avoid any form of plagiarism.*

Nobody preparing a publication should use the results of other researchers without correctly attributing them. The worst 'crime' that can be committed against the code of honour, is to publish a paper in which results are stolen from another source, which is not cited, and published as the author's own.

▷ *Authorship should be limited to those who have made a significant contribution to the concept, design, execution or interpretation of the research.*

Honorary authorship or coauthorship by people who have not contributed to the research are to be frowned upon. Such authors may be included because they are famous, and thus expected to give the paper an easier passage throught the refereeing. Or they may be included out of gratitude, e.g. the head of department who played a vital role in getting the project funded in the first place. Whilst one can generously consider the latter as a contribution to the overall concept of the research, the former is clearly a form of cheating that should not be tolerated.

In one extreme case, a scientist who served largely as administrator for a research group, and himself performed no research, persuaded the group members to list him as last coauthor on all their publications, thus generating the impression that he was actively involved in the group's research. After several such papers had been published, he collected these together and submitted them as a '*Habilitation*' thesis (the degree, somewhat higher than a PhD, required to become a university lecturer in Germany). Surprisingly, perhaps, he was awarded the degree. Yes, occasionally crime pays, but is it a risk you would want to take? The examples in the next section will show how easily such scientific misconduct can ruin a scientist's career for good.

▷ *All the coauthors share some degree of responsibility for the content of the papers on which their names appear.*

The content of a scientific paper is the responsibility not only of the principal author, but also of those coauthors who have provided scientific leadership for more junior colleagues. Coauthors who have made only limited contributions (e.g. to the interpretation of the results but not to the experiment) are responsible only for the parts of the argument that they contribute. Nonetheless they are requested to check the whole content to ascertain, as far as is possible, that it is valid and accurate. Any author who is not willing to accept appropriate responsibility for the contents of a paper should not agree to be coauthor. This paragraph was added to the code following the scandal, to be described below, surrounding the papers of J.H. Schoen.

4.3.1 Historical Cases of Unethical Scientific or Publishing Practice

Despite the essentially noble aims of science, scientists are still human and are just as susceptible as the rest of society to the temptations to cheat for reasons of personal gain.

The history of science reveals plenty of cases of intentional deception, as well as instances where sloppy practice has led to the publication of sensational, but incorrect, results. A recent high-profile case, in which a young researcher brought disgrace upon himself, his colleagues and the institutes in which he worked, is that of Jan Hendrik Schön.

Jan Hendrik Schön

In February 2000, Schön, a promising young physicist, published some startling experimental results. Schön and his partners, working at Lucent Technology's Bell Labs and at ETH in Zurich, had started with molecules that don't ordinarily conduct electricity, and claimed they had succeeded in making them behave like semiconductors. These findings were reported in the flagship journal Science and created an immediate stir.

Several further papers followed in quick succession, many of them published in Science or Nature. Schön's group reported that they could make other nonconductors into semiconductors, lasers and light-absorbing devices. These claims were revolutionary and their implications for electronics and computing enormous, holding the promise of flexible plastic electronics and vastly miniaturized computer components. As one Princeton professor stated, Schön had 'defeated chemistry'. Schön continued to churn out the articles, publishing more than 90 in about three years. In 2001, he received an award for the scientific 'Breakthrough of the Year', but most scientists saw this recognition as only the beginning and already viewed him as a contender for a future Nobel Prize.

But suddenly, everything went horribly wrong for the young Wunderkind. In April 2002, a small group of researchers at Bell Labs contacted Princeton physics professor Lydia Sohn and voiced their fears that all was not right with Schön's data. Sohn recalls that she and Cornell University's Paul McEuen stayed up late one night and found some disturbing coincidences in Schön's results: Plots used to illustrate the outcomes of completely different experiments displayed identical background noise, and some even showed the same data points. *'You would expect differences,'* she said, *'but the figures were identical. It was a smoking gun.'*

His suspicions aroused, McEuen started looking closely at Schön's other published work. He found many duplicate graphs in different papers on different subjects. Schön was apparently using the same sets of pictures to tell lots of different stories.

In May, McEuen and Sohn formally alerted the editors of Science and Nature to the discrepancies. McEuen and Sohn also informed Schön, his supervisor and coauthor, and Bell Labs management that they were blowing the whistle. Schön immediately insisted that his experiments were fine, and that the duplicated figures were a simple clerical error for which he now offered substitutes. To Nature he declared he was *'confident'* of his results. To Science he said, *'I haven't done anything wrong.'*

The end of the story is that Jan Hendrik Schön was fired from his job at Bell Labs after an independent investigatory committee, headed by Stanford University physics professor Malcolm R. Beasley, concluded that he had fabricated the spectacular findings he had reported in the field of molecular electronics. He was found guilty on 16 out of 24 charges of scientific misconduct.

In a response to the committee's report, Schön admitted making *'mistakes,'* but said he did not intend to mislead anyone. *'I have observed experimentally the various physical effects reported in these publications, such as the Quantum Hall effect, superconductivity in various materials, lasing, or gate-modulation in self-assembled monolayers, and I am convinced that they are real, although I could not prove this to the investigation committee.'*

Schön's response did not convince the authorities. Not only did he suffer the ignominy of the much publicized scandal: Prizes that he had been awarded were withdrawn and he lost both his existing job and his prospective post as director at the Max Planck Institute in Stuttgart. It is unlikely that he will ever be able to return to work as a scientist. Even if some institute were willing to employ him, who would believe his results?

At the time of this debacle, there were an estimated 100 laboratory groups working on Schön's results in the United States and around the world. For graduate students basing their Ph.D. research on Schön's experiments, their education was suddenly at stake. Postdoctoral fellows had to worry about their prospects for future employment. Some junior professors had even tied their bids for tenure to experiments based on Schön's findings. So, beyond ruining his own career, Schön also put the professional futures of many of his scientific colleagues at risk.

Other Selected Cases of Scientific Fraud

1912: Piltdown Man

In Piltdown, Sussex, archaeologist Charles Dawson unearthed two ancient humanoid skulls and claimed that 'Piltdown Man' proves humanity originated in the UK after all. It wasn't until 1953 that it was proven that 'The First Englishman' was just a dull, medieval skull, with the jaw of an Orangutan attached to it.

1926: Fabricated Frogs

Hailed as the new Darwin, the brilliant biologist Paul Kammerer saw his career come to an end when Nature magazine accused him of tinkering with pictures of evolving frogs. The accusation is still unproven, but Kammerer was so severely discredited that he put a bullet in his head.

1989: Cold Fusion

On March 23, 1989, Stanley Pons and Martin Fleischmann announced their discovery of '*cold fusion.*' It was the most heavily hyped science story of the decade, but the awed excitement quickly evaporated amid accusations of fraud and incompetence. When it was over, Pons and Fleischmann were humiliated by the scientific establishment; their reputations ruined. They fled from their laboratory and dropped out of sight. '*Cold fusion*' and '*hoax*' became synonymous in most people's minds, and today nearly everyone believes that the idea has been discredited.

1997: German Scam

Germany was shocked to learn that two prominent cancer researchers, Marion Brach and Friedhelm Herrmann, had been concocting research results for years. In perhaps the biggest scientific fraud scandal in Europe to date, the two researchers were accused of faking data in at least 12 publications.

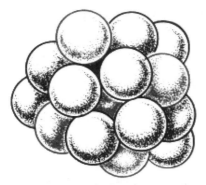

Fig. 4.8. Atomic nucleus. The race to identify ever-heavier elements has tempted at least one scientist to publish tesults that were wishful thinking

2000: Caught In The Act

A respected Japanese archaeologist, Shinichi Fujimora, was caught on film while burying early Paleolithic relics, stone tools that he had discovered elsewhere, at the site of one of his digs. Evidently he intended to allow these to be 'found' at the new sites, enabling him to claim for them an historical importance that they did not really possess. After being caught red-handed, Fujimora was fired. All of his previous archeological work came under suspicion, and Japanese history books had to be rewritten.

2002: Elements Number 116 and 118

California's prestigious Lawrence Berkeley National Laboratory woefully admitted that one of its scientists had fabricated the discovery of two new chemical elements: elements 116 and 118. The researcher, Bulgarian Victor Ninov, who was in charge of the data processing, was summarily dismissed. The embarrassing stain remains, however. How could a sole researcher fake the discovery of something so massively important as two new elements?

What Motivates Scientific Misconduct?

Whilst many the above cases of falsification are clearly based on intentional fraud, there is also a huge grey area, which one might characterize as sloppy work motivated by wishful thinking. Researchers are often so convinced that their pet theory is correct that they become blind to any data that does not fit neatly into the picture. This is a particular danger in fields where experiments are not easily reproducible, for example, in the field of cold fusion, which actually continues – despite its unhappy beginnings – to attract occasional scientific attention on the fringes of mainstream research.

Furthermore, there is often a conflict of interest when it comes to publishing results. Scientists are under pressure to publish, for the sake of their careers. And – in a field where there is a lot of competition – they often feel pressed to publish fast, or risk someone else getting the credit for the idea. Thus the temptation to ignore uncomfortable contradictory results is large and some even succumb to the desire to fill in gaps with invented results. As the above cases show, however, the fame that one gains through such practices may not be of the desired kind. The best policy is always to be honest. If you are in a hurry to publish, at least be honest about the shortcomings of your results and draw attention to any gaps or open questions. That way, no one is likely to accuse you of malpractice, even if it does turn out that your interpretation is wrong. A hard-earned scientific reputation, built up over years of careful and patient work, can be destroyed overnight by intentional fraud, and put severely at risk by exaggerated claims, sloppy work or half truths.

4.3.2 Gatekeepers

Who is responsible for maintaining the ethical standards in scientific publishing? First and foremost, of course, the scientific enterprise has to rely on the honesty of individual scientists. Thus the main responsibility for published claims lies with the leading author who carried out the bulk of the work/analysis. Senior coauthors with the function of mentor (typically PhD supervisor, head of department or responsible professor) must also take responsibility for filtering out anything that smacks of misconduct.

A sensible practice that is adopted in many scientific departments is the requirement that every paper be checked by the re-

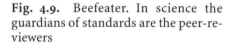
Fig. 4.9. Beefeater. In science the guardians of standards are the peer-reviewers

sponsible professor or head of department prior to its being submitted for publication. This also helps to filter out papers that are simply superfluous or badly written, and, in the long run, helps the department to gain and maintain a reputation for excellence.

Of course the journal publishers themselves also endeavour to exercise a kind of policing function. Journals usually have editorial boards, whose members either referee the papers themselves or pass them on to qualified peer reviewers to check. It is the task of the referees to evaluate the quality of the work reported. Does it make sense, scientifically? Are the conclusions borne out by the results presented? Is there sufficient new information to warrant publication in the journal in question? Do the authors adequately describe the current state of knowledge in the field and give the appropriate references to others who have contributed?

Although some cases of plagiarism or dubious scientific practice might be caught by referees, they rarely need to ask themselves 'Are these data really believable, or could they have been fabricated?' Indeed, there is usually no way for them to check this anyway. Of the dozens of referees who checked the Schön papers submitted to Nature and Science, none even hinted that the data might have been created rather then measured. The results, unique as they were, nonetheless sounded feasible. Even plagiarism is often hard to catch. However compentent the referees are, it is not feasible to expect them to have read and remembered every article published on the topic at hand.

The work of referees makes a vital contribution to maintaining scientific standards. Thus it should not be taken lightly. Referees who are sloppy themselves, or who misuse their privileged position are also guilty of scientific misconduct. If, as a referee, you are asked to evaluate a paper that lies outside your sphere of expertise, you are obliged to pass it back to the sender, or – as sometimes permitted – to pass it on to another colleague who is more familiar with the area concerned.

Or what if you are in direct competition with the authors of a paper you are asked to referee? Well, there are two ethical choices and one unethical one. You can review the paper as quickly and fairly as possible, i.e. do your duty without regard to the competitive element. Or you can send the paper back immediately to the journal explaining that there is a conflict of interest and another referee should be chosen. The third option, a case of which is known to the present authors, is an example of severe refereeing malpractice. In this particular case a referee intentionally delayed the publication of an interesting paper by demanding time-consuming revisions. In the meantime, his group quickly put together their own paper on the topic – using insights gained from the paper to be refereed – and managed to get their paper published first. Although this does not officially give them priority (which is determined by date of submission), it nonetheless allowed them to make the big bang and receive the acclaim.

4.3.3 Conclusion

Despite what might seem like a long list of frauds, misdemeanours and potential temptations, science is an endeavour which is overwhelmingly characterized by honesty. The vast majority of scientists enter the profession because they are truly interested in discovering new phenomena and gaining new insight into the workings of nature. Although not immune to the human failings of self-interest, ambition and greed, most scientists find more satisfaction in pursuing the truth that in indulging these baser instincts. This is likely to preserve science in the long run and ensure that the scientific enterprise, in contrast to business or politics, can continue to function on the basis of mutual trust and respect.

5 Writing and Publishing Your Own Paper

In this chapter you will learn when and how to plan, prepare and submit a scientific paper reporting the results of your research work. If you have never done this before, it is advisable to read all sections of this chapter carefully. For those who have already published papers, this summary of good and bad practice will still be useful for improving your writing habits.

Although we will concentrate here on the writing of papers rather than books, you will find that once you have mastered the skills needed to prepare a good paper, you are already half way to preparing good teaching material and even book manuscripts.

5.1 Planning and Preparation

'Failure to prepare is preparing to fail.'

Mike Murdock

5.1.1 Before You Begin

Before even planning a publication, it is advisable to ask yourself the following critical questions:

Do I Have Enough Significant Results? Weigh the value of your results critically and carefully before deciding whether they are worthy of publication. In other words, ask yourself: Are my results truly reproducible? Do they represent an essential step forward? Will they be of interest to other scientists? If you can answer *'yes'* to all these questions, then a publication is probably justified. On the other hand, if your results contain little new information, or require further supporting results before they can be interpreted, you should consider waiting until you have something more substantial to report. Usually your supervisor or head of department can advise you on this point. Discuss your results with him or her in the context of other recent work in the field. This will help to establish whether your contribution is really going to fill a gap, settle a controversy, or even strike out in an entirely new direction.

What Do I Want to Report? A good publication usually concentrates on a restricted set of important results and the message that they contain. Thus you should begin by deciding on the message that your paper will convey and then choose the results that are necessary to support this message. Overloading the publication with other, less relevant results, should be avoided. Your paper should be as focused as possible; that way, the reader will be more likely to appreciate and remember it.

When Should the Results Be Published? Occasionally, there are good reasons for delaying a publication. As already mentioned, it is important to wait until you have enough significant results and have checked their reproducibility. But there is also another reason: When scientific results have interesting technological implications or possible commercial exploitation, you should consider applying for a patent to cover the relevant applications. In Europe and Japan, at least, an idea that has been published can no longer be patented. In Chap. 7 we give a short introduction to patents and patent applications.

Am I in a Hurry? If many people are working on the same, or a similar, topic a race may develop to answer certain questions and publish certain key results. Several groups may be attempting to reach the finishing line with the first seminal publication.

In this case it is natural, as soon as you have some significant results, to publish them as quickly as possible so as to gain priority of discovery for yourself and your co-authors. A short note or letter is then often the appropriate type of paper. Such a paper, however, cannot give anything more than a rather superficial discussion, since it is limited in length. Therefore, if you are not about to be pre-empted by other researchers and your results lend themselves to publication as part of a larger and more detailed study, it may be better to take more time and prepare instead a longer original paper. It is a shame to 'waste' good results by publishing them hastily in a sloppy publication rather than exploiting them to the full in a well argued and highly acclaimed paper. You may have to invest more time in collecting further data to complete the picture, in thinking carefully about the theoretical interpretation, and in searching the literature to provide sufficient background; but in the end a substantial paper may have more impact than a threadbare letter or short note.

One can also wait too long: Carbon nanotubes were in fact first discovered by Mr. Iijima of NEC, Tsukuba in 1990. But he did not publish his findings until two years later. Although they were found worthy of a publication in Nature (REF), this did not change the fact that he was too late to be considered for the related Nobel Prize awarded for work on fullerenes.

In some cases, however, publishing too early can be as bad as publishing too late. If you are too late with your paper, others will have stolen the limelight, and the interest in your findings – even if they are slightly different to those already published – will be diminished. But it can also happen that your work is ahead of its time. In other words, you may be working in a field that is yet to become fashionable or will, in future, gain in technological relevance. There is not much you can do about this, as one of the present authors discovered to his disadvantage. Later researchers will not necessarily rediscover your 'pioneering' publications, and will only cite recent mainstream publications on the topic. Here one simply needs a good nose for being in the right place at the right time!

What Is Already Known? Although you will surely have done some background reading before you started your investigation, the outcome of your study may necessitate further reading before it can be fully interpreted and set into the greater scheme of things. Other publications may be useful in supporting your interpretation; or they may cast doubt on it.

Which Language? Your publication will reach a wider audience and have more impact if you publish in an international English-language journal. If English is not your native language, then it may cost you a much greater effort to prepare your paper in English. Furthermore, the chance of the paper being accepted by a national journal is probably higher. But, if your results are really worth reporting, then they should be made known to the interna-

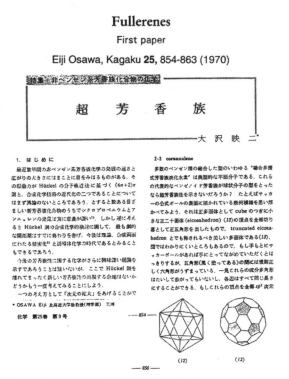

Fig. 5.1. Eiji Osawa's paper on fullerenes in the Japanese-language journal Kagaku [25, 843 (1970)] failed to attract international attention

tional community. The serious drawbacks of not publishing in English have been experienced by a number of Japanese researchers: The most famous example is probably Eiji Osawa of Hokkaido University who first predicted the existence of fullerenes in 1970 (although he did not use the same language to describe them; he simply predicted the existence of this new modification of carbon that was so far unknown). His paper is shown in Fig. 5.1. When the Nobel Prize was awarded for the discovery of fullerenes, he was not among the winners. Osawa had published his results in a Japanese-language journal, with the consequence that he remained virtually unknown internationally and his paper was not accessible to the Nobel Prize committee. Attempts to have the decision changed or to add a further prize-winner after the event were unsuccessful. After he had sent his paper to the Nobel Prize Committee, he was told, in as many words: '*Sorry. The rules for awarding the Nobel Prize say that it can be awarded to at most three researchers. We cannot add a fourth name. And the decisions of the Nobel Prize Committee are never changed.*'

5.1.2 Choosing a Journal

It is advisable to decide on a journal in advance, since each journal has its own particular style, scope and rules. It will help you to get your paper accepted if it is written with the particular journal in mind.

New or Established Journal? If you choose a new or little-known journal that is begging for papers, you will have a good chance of your paper being accepted. But new and lesser-known journals are not so widely read as established journals and your paper will have a lower impact, and correspondingly fewer citations.

An acquaintance of the authors was persuaded to publish some of his work in a newly founded journal. This article is, to date, the most cited article ever to have appeared in this journal. The author was happy until he spoke to the editor of another, better known, journal in the field. Unimpressed, he remarked: '*Yes, but if you had published this important paper in our journal, it would have got ten times more citations*'.

Rejection Rates. If you submit your paper to a leading and highly cited journal, you must be aware that there is a higher probablility that it will be rejected; and even if accepted, you may well be asked to make careful revisions to satisfy the referees. Established journals naturally receive more papers and can afford to have a higher rate of rejection. For example Physics Review Letters rejects about 70 % of the papers submitted, and for Nature the rejection rate is close to 90 %. But if you are confident of having done good work, this should not deter you. Even these prestigeous journals need to fill their pages!

Other Factors. You will already have decided whether to write a short note or a longer paper. This may affect your choice of journal, since some journals publish exclusively short notes/letters, others a mixture of letters and longer papers, whereas still others have only longer papers and/or reviews.

Another decision you will need to make is whether to aim for a specialist journal, such as '*Progress in Statistical Hydrodynamics*', or for a journal covering the whole of your discipline (e.g. Physical Review Letters for physics), or even the whole of Science (Nature and Science are the two main pan-scientific journals). You should only choose such general journals if your findings are so significant and so exciting that they will be of interest outside your own field.

5.1.3 Who Should Be Named as Authors?

Number of Authors. Often the question of who should be included among the authors has a simple and clear-cut answer. The paper should be authored jointly by those people who made a substantial contribution to carrying out the work reported, including the theoretical interpretation – if any – of the experimental results. In other cases the answer is not so simple. What about the guy who set up the experiment originally, but was not involved in taking data from your sample? Or the technician who helped to fix the equipment when the pump exploded? Or the head of the research group, who suggested the experiment in the first place, but had no time to follow its progress? Or the colleague who helped to solve that impossible integral?

Usually, it is better to keep the number of authors to a minimum and for these to be the people who have done the work from beginning to end and can explain everything about it if required. The others whose assistance was indispensable, should be appropriately acknowledged at the end of the paper.

In some institutes and some fields there are special rules. Sometimes the head of an institute or research group co-authors all papers written by members of his staff. In particle physics, where huge groups of people frequently collaborate in a major project at an accelerator, it is common to include all of them on every paper; hence papers with a hundred or so authors are not unusual.

For some sponsored research projects it is a condition that the project director is listed as a coauthor if a publication is to count as a report of project results. And if your research group is hoping to receive a grant from the same source for their next project, then it is wise to oblige!

Sequence of Authors. Having decided who will be named as authors, the next most delicate question is in what order the names should appear. It is a widely accepted and fair rule that the first author should be the person who made the greatest contribution to the work. This of courses begs the question of how to evaluate contibutions; should one choose the person who had the original idea, the one who did most to realize the experiment; the one who spent the most hours in the lab? This can be a tricky decision and sometimes – to avoid unpleasant argument – it is better to leave the final decision to the head of the group, whether or not this person is themselves an author.

In some disciplines and certain institutes other rules are used. At some university clinics the director is always named as the first author, regardless of his contribution. In other cases, senior staff members make use of their status to insist on being named first and, sometimes, they also write or rewrite the papers. This may or may not make for a better paper, but it does not really help the junior colleagues to learn, nor does it give them due credit. A more generous leading scientist or institute director will agree with the statement of one of our acquaintances: *'The main work is in the production of the scientific results and not in their presentation. Thus I give the credit to the person who did most of the work. It does not matter that insiders may notice who wrote the paper.'*

Some institutes insist on authors appearing in alphabetical order. But this can severely disadvantage key authors whose names fall at the end of the alphabet: They may have had nearly all the ideas and done most of the work; but if they become the last of several authors, their name may rarely be connected with the paper since they will frequently become an anonymous part of the '*et al*.'.

Faced with this problem, an ambitious associate professor at the University of California actually changed her name, from Yard to Ard, so that she could appear as first rather than last author on her publications.

If such rules are observed in your group, or at your institute, it is wise to accept them. Otherwise, aim for the minimum number or authors consistent with doing justice to the main contributors to the work, and order these authors according to the extent of their contibution.

Ghost Authors. It is not unknown for renowned scientists to be asked to appear as co-author of a paper written by junior colleagues, even though he or she had little or nothing to do with the work reported. This is done in the belief that a *'big name'* will help the paper to be accepted for publication. Other types of wheeling and dealing include bribery to become an author, bribery not to be an author (e.g., if someone is particularly keen to be the first author). None of this is a good idea: Sooner or later it may cause severe problems, especially if practiced repeatedly. With your scientific reputation at stake, it is better not to contemplate any such manoeuvring.

5.1.4 Coordination Between Authors

Note that being the author of a paper does not necessarily imply that you need be involved in writing it. However, each author should have had the opportunity to read and approved the contents of the text. But who should actually write the paper? Should each author be given a section to write? This, in fact, is probably the worst solution; the result will be incoherent in content and style and the process reminiscent of constructing the Tower of Babel.

The best approach is to select one person to do the writing initially (occasionally two, if there are theoretical and experimental parts to the paper). When the first draft is finished it should be given to all authors for their comments and criticism. Step by step they give their input and, after a few iterations, the result is a coherent paper that all authors are happy with.

But now we are getting ahead of ourselves. Having decided who will actually author the paper, the next step is to plan its contents in detail.

5.1.5 Planning and Preparing the Contents

Faced with the task of writing a scientific paper, especially if it is their first, many people break out in a cold sweat and have no idea of how to begin. They dig out some data, compare it with other data, write a few lines about it, read a related publication and cite it, write a few more lines about the experiment, scratch their heads to remember where they put the data from the control experiment ... This style of writing might be called the *'let's see where we end up'* approach to preparing a paper.

It is far, far better to decide where you are going before you set out on the journey. The plan, and even the destination, may change as you progress, or because new ideas or results become available, but you should always have a clear plan in mind.

From our own experience in writing papers and also from discussions with colleagues and friends who are experienced and successful scientists and scientific writers, we would like to propose an optimal approach to drafting a paper. It consists of several steps, which should be tackled in turn. Before you start, all you need is a fairly clear idea of which experimental or theoretical topic will be addressed. Note that for papers containing only text and/or mathematics (i.e. a minority of theoretical papers only), the preparation procedure will begin at stage III, rather than stage I.

I. Assemble Data/Results

Collect together and analyse all relevant data, plots, and theoretical results that are available. This is the raw material with which you will work. In the process of collecting and analysing this material you might already notice one or more gaps. If these are critical, then – if possible – go back to the laboratory or computer and acquire the missing information.

II. Choose/Prepare Figures

You will see from the large pile of paper and/or the numerous electronic files that not all data can be included in your publication. The second step is to select the most important results and use these to plan and draft the figures for your paper. Many experienced scientists recommend preparing the figures and tables of a paper before writing the text. In an experimental paper, at least, the figures often convey as much, or more, information as the text. So do not underestimate their importance or simply regard them as a decorative extra. If you will be presenting data in graphical form, it is wise to decide now which parameters are to be correlated with one another and also the best method of plotting the data: What makes the effects you have observed clearest: linear, logarithmic, or half-logarithmic plots? Pie charts? Bar charts? Contour plots? As in

Volume 45, Number 6   PHYSICAL REVIEW LETTERS   11 August 1980

New Method for High-Accuracy Determination of the Fine-Structure Constant Based on Quantized Hall Resistance

K. v. Klitzing
Physikalisches Institut der Universität Würzburg, D-8700 Würzburg, Federal Republic of Germany, and Hochfeld-Magnetlabor des Max-Planck-Instituts für Festkörperforschung, F-38042 Grenoble, France

and

G. Dorda
Forschungslaboratorien der Siemens AG, D-8000 München, Federal Republic of Germany

and

M. Pepper
Cavendish Laboratory, Cambridge CB3 0HE, United Kingdom
(Received 30 May 1980)

Measurements of the Hall voltage of a two-dimensional electron gas, realized with a silicon metal-oxide-semiconductor field-effect transistor, show that the Hall resistance at particular, experimentally well-defined surface carrier concentrations has fixed values which depend only on the fine-structure constant and speed of light, and is insensitive to the geometry of the device. Preliminary data are reported.

PACS numbers: 73.25.+i, 06.20.Jr, 72.20.My, 73.40.Qv

In this paper we report a new, potentially high-accuracy method for determining the fine-structure constant, α. The new approach is based on the fact that the degenerate electron gas in the inversion layer of a MOSFET (metal-oxide-semiconductor field-effect transistor) is fully quantized when the transistor is operated at helium temperatures and in a strong magnetic field of order 15 T.[1] The inset in Fig. 1 shows a schematic diagram of a typical MOSFET device used in this work. The electric field perpendicular to the surface (gate field) produces subbands for the motion normal to the semiconductor-oxide interface, and the magnetic field produces Landau quantization of motion parallel to the interface. The density of states $D(E)$ consists of broadened δ functions[2]; minimal overlap is achieved if the magnetic field is sufficiently high. The number of states, N_L, within each Landau level is given by

$$N_L = eB/h, \qquad (1$$

where we exclude the spin and valley degeneracies. If the density of states at the Fermi energy, $N(E_F)$, is zero, an inversion layer carrier cannot be scattered, and the center of the cyclotron orbit drifts in the direction perpendicular to the electric and magnetic field. If $N(E_F)$ is finite but small, an arbitrarily small rate of scattering cannot occur and localization produced by the long lifetime is the same as a zero scattering rate, i.e., the same absence of current-carrying states occurs.[3] Thus, when the Fermi level is between

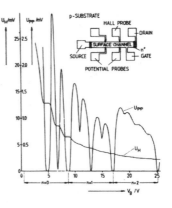

FIG. 1. Recordings of the Hall voltage U_H and the voltage drop between the potential probes, U_{pp}, as a function of the gate voltage V_g at $T = 1.5$ K. The constant magnetic field (B) is 18 T and the source drain current, I, is 1 μA. The inset shows a top view of the device with a length of $L = 400$ μm, a width of $W = 50$ μm, and a distance between the potential probes of $L_{pp} = 130$ μm.

494

Fig. 5.2. Key figure in a paper that led to a Nobel Prize

a presentation, beware of overloading the reader. A figure should only try to convey one essential idea.

You might already try to choose one figure as the key figure of the paper. Many seminal papers gain their fame as a result of one key figure demonstrating an important effect (Fig. 5.2).

Then you should write short captions for each figure, describing what they show, but with no discussion.

Finally, you should arrange the figures in the appropriate order. This yields a set of anchor points that define the main content and provide a basis for the logical arguments that are given in the paper.

III. Define the Main Message

As a third step, you might ask yourself: What is the essential new information and what do we want to conclude from it? Prepare a first draft of the conclusions. By determining the endpoint in advance, you will have a clearer idea of where the remainder of the text has to lead. Of course you can modify them later, but it is very useful to have a preliminary set of conclusions in mind as you write the rest of the text.

IV. Construct a Skeleton

The rough planning can now be taken a stage further. Write an outline of the paper, defining the various sections it will have (see Sect. 5.3), and making a list of key points for the descriptive and discursive parts. This plan, too, may well change during the course of writing, but such a skeleton will help you to organise your arguments in a logical order. All coauthors should have the opportunity to approve this outline, also the choice of figures, and especially the conclusions.

V. Assemble Literature

You will have presumably already done all the necessary background reading on the topic of your paper. You will also have kept up to date on the latest advances made by other researchers. So all that should be necessary is to collect all the papers to which you will want to refer. These will typically be: (1) One or more older seminal works that gave rise to the field of research, cited in the introduction. (2) One or more review papers on the topic, also cited in the introduction so that interested newcomers to the field can find the background information themselves. (3) Closely related

Fig. 5.3. It is good to collect and refer to the relevant literature, but keep the pile manageable!

research reports that left open the questions you have tackled; contradicted one another, hence inspiring your experiment to settle the question; presented results that you have made use of in your analysis; or could be useful to the reader of your paper by supplementing the information you give (see also Sect. 5.3.9.) In principle it is advisable to have read – and to have available – all the papers you cite. There is a widespread habit of copying citations of older papers from reviews. Whilst this surely pleases the authors of the publications concerned, it does not necessarily benefit the reader of your paper, who wastes time searching for papers that turn out not to be relevant.

Our personal recommendation is to have available hard copies of all the papers you cite. Sort these papers into the sequence of citation in accordance with the order of your figures and the skeleton contents.

VI. Write the Text

You should now have in front of you, in the correct sequence, all the material needed to write the paper. You will be surprised to find that a large part of the work is already done! What most people imagine to be the hardest part of the job – writing the text – turns out to be simply a matter of filling in the gaps. In Sects. 5.2–5.4 we will consider the questions of good scientific writing style, how to structure your writing, and also describe the main formal rules in scientific writing.

Fig. 5.4. Writing the text is not as difficult as most people imagine

VII. Refine/Revise the Text

When you have started writing and are trying to present every-thing in a coherent and convincing way, you will often find some of your arguments are not as clear as you initially thought. There are still some gaps requiring you to do further reading, thinking, talk-ing to colleagues and, in the most extreme case, additional experi-ments or calculations. But provided everything else is well organ-ized, you will also master these problems.

Discovering new gaps that you had not previously noticed is a common phenomenon that arises when you deliberate intensely about the presentation of your results. In doing so you come to un-derstand them better yourself. When you are obliged to describe your findings verbally and in detail, be it in a paper, a talk or a tu-torial for students, you will quickly discover whether your descrip-tion is easily understandable to other scientists. As you tackle this task, you will often arrive at new insights, perceive new connec-tions, or, indeed, discover unanswered questions that need to be followed up in future work. Thus, in common with teaching, the preparation of a publication can turn out to be very instructive, since the mere process of organizing ones thoughts can con-tribute – often unexpectedly – to the creative process.

Eventually, you will have completed a first draft that you are content with. If the paper has coauthors, now is the time to seek their final approval of the text. Usually they will suggest at least minor revisions, and several iterations may be necessary before all coauthors are satisfied. At this stage, you might also take the opportunity to ask experienced colleagues who were not directly involved in the work to read and comment on the paper.

5.2 Style of Writing

5.2.1 Who Will Read Your Paper, and Where?

Before starting to write the text in earnest, it is important to have in mind what readership you are addressing. This will have an influence on your style of writing and, in particular, on the amount of background information that you will have to provide. The interest in your paper will be far greater if you pitch the discussion at a level at which the readers feel at home. Hand in hand with the question of readership goes the matter of which journal you propose to publish in. There are differences in the style of journals that you should be aware of. Specialist journals, addressed only to the experts will expect a higher-level and more technical discussion than a periodical aimed at a more general audience, for example Science or Nature. Here your article will probably be read by non-specialists too, and should thus provide the necessary background information to give them access to the forefront research described.

5.2.2 Keep It Clear, Concise and Logical

Regardless of which journal/readership you are aiming at, you should always try to write as clearly as possible. If every second sentence contains references to obscure theorems, laws and results that need to be looked up elsewhere, then even specialist readers may quickly become frustrated, at best jumping to the conclusions to check whether there is really anything important buried among the frightening detail. For the same reason, it is advisable to avoid using jargon; and acronyms, except the most widely used ones, should be explained.

The description and discussion of your work should be logical and concise. Historical descriptions of the field should be kept to a minimum. Some brief words to indicate why your research is important in a larger context should suffice. The style should also be didactic, in the sense that the reader is systematically guided through the argument and does not have to make giant leaps, of calculation or of faith, along the way.

5.2.3 Scientific Prose

In a scientific paper the style of writing is simpler, but more formal than that encountered in other books, newspapers, or everyday speech. This style has arisen over time as a by-product of the scientific method itself. Corresponding to the logical approach to scientific discovery, scientific writing must likewise be logical, clear and unambiguous. A large vocabulary is not essential, since scientific terms are generally well defined and flowery descriptions, of the type found in novels, are to be strictly avoided. If you are not writing in your native language, it is advisable to check with colleagues and dictionaries on the meaning of any terms you are unsure about. Familiarity with the existing literature in your field will help you both with vocabulary and style of writing. It is wise to show a draft of your paper to colleagues and to ask for their comments on both the content and the style of presentation. Having to listen to the critical remarks of well-intentioned colleagues is preferable to receiving damning referee reports or – should it slip through – to having your important results presented in an embarrassingly badly written publication.

Unless you are really confident in expressing yourself, it is always better to err on the side of simplicity. Stick to short sentences and straightforward descriptions. Present the information in a logical order and avoid getting sidetracked into unnecessary detail or discussion.

In general an impersonal style is to be preferred: Thus, rather than '*I found ...*' it is better to write '*we found ...*' or even '*one finds ...*', the latter reflecting the fact that the results should be reproducible, no matter who is doing the investigation. Purists would even insist on a passive style of presentation, in this case: '*It is found that ...*'. This passive form of expression is actually tantamount to taking all human involvement out of the picture. This, we feel, is overdoing the aim to appear objective. It is fine to use the passive voice where appropriate, but by no means essential to use it everywhere.

5.2.4 Importance of Good English

If English is not your native language but you wish to publish your work in an international English-language journal, then you face an additional challenge. It is essential that the standard of the English in your paper is at least adequate to convey the intended information to the reader.

From our experience in processing articles for journals, inlcuding much feedback from referees, we observe that language problems are especially prevalent in papers written by Chinese, Russian and Japanese authors. Although these papers may be dealing with good research work, it is not uncommon to find that they are rejected by referees for the simple reason that it is impossible to understand what the author is trying to say.

Thus all non-native speakers are encouraged to have their papers checked by colleagues who speak at least reasonable English. In the past, I (CA) have often received papers by Russian authors to referee for journals. The editor typically said: '*You learned Russian. Perhaps this will help you to understand what these authors are trying to tell us.*' It is true that one can often decipher the meaning of the strange English if one knows the author's native tongue. I occasionally helped to rewrite promising papers submitted by Russians. But most referees don't have the time and energy for rewriting papers. If they are sympathetic to the author's attempt, they will ask him or her to rewrite the paper in better English. But more often than not, the paper will be summarily rejected.

Is there anything that non-native speakers can do, apart from taking a crash course in English, to improve their papers and avoid confusing the referees? When I (CA) made my first attempts to write papers in English, my style of writing was severely criticized by American and British referees for being 'very German'. This, it turned out, meant that the sentences were too long, too many clauses were nested within sentences, and of course the choice of vocabulary was influenced by the German usage. The rules that I constructed for myself in order to write better English language papers were:

▷ Keep most sentences short, preferably no longer than 15–20 words.

▷ Avoid using many linked or nested clauses.

▷ Restrict each sentence to one idea.

▷ Keep to an active style for the mostpart; it is simpler to write.

▷ Find out about, and try to avoid, the typical recurring errors made in English by speakers of your language (e.g. Germans often use verbs as nouns; Japanese confusion of the letters 'l' and 'r'; Russian tendency to omit articles where they are needed.)

Fig. 5.5. Pay attention to style of writing

At the end of the day, good scientific writing style comes largely with practice. To speed up the learning process, it is advisable to read plenty of good papers, paying attention to their style, and – for your first few papers at least – to ask experienced colleagues for constructive criticism.

5.3 Structure of a Scientific Paper

In school science classes most of us will have learned about the conventional structure of a scientific report: It begins with an '*Introduction*', followed by the further sections entitled, in turn, '*Method*', '*Results*', '*Discussion*', and finally '*Conclusion*'. This simple and logical sequence of information has – over the past century and more – become well-established as the internationally accepted structure for reporting on scientific findings.

But this structure actually reflects a way of thinking that, initially at least, was common only in the western world.

Western Style
The Western style of writing a scientific report can be likened to a tree. It is very hierarchic, a logical sequence of explanations, each based on the one before, not leaving anything unexplained before discussing it. Starting from a general and broader point of view, the trunk, the discussion progresses in a number of directions, the branches, at the same time becoming more detailed. Finally, the endpoints of the various threads of discussion are drawn together in an overall conclusion.

Traditional Japanese Style
As a visiting professor at Kyoto University in Japan, one of the authors (CA) was frequently asked to check, and often rewrite, the papers from his institute. This turned out to be very revealing, showing that the Japanese way of thinking and presenting their findings is often quite different. They attach greater importance to the details. This is also reflected in other aspects of Japanese cul-

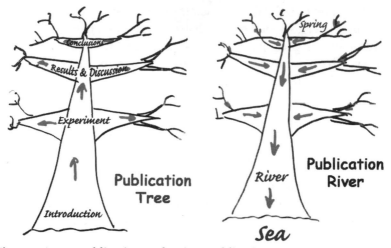

Fig. 5.6. A tree publication and a river publication

ture, e.g. the way they wrap presents or present food. In a sense it is the exact opposite of the Western '*tree*' (Fig. 5.6). Instead it may by likened to a river, rising from numerous small sources, each of which forms an ever stronger tributary, these streams all flowing into the main river and helping it to swell. Thus this approach, shown schematically in Fig. 5.6, begins with a set of details, which may even include the conclusions, and works backward from these to construct the main picture, which in turn justifies the original assumptions. In other words, in the Japanese way of thinking, first all the information is gathered and then, at the end, the overall picture is developed.

This difference in logical structure goes hand in hand with the fact that the traditional way of thinking in Japan is much less abstract than in the West (of course this generalization does not hold for all individual Japanese scientists).

Recommended Structure and Style

The structure that will be described in the following sections is essentially the one given at the start of this section. But in contrast to the simple structure school pupils are taught, research publica-

Fig. 5.7. The importance of structure: Stone garden of a Zen temple

tions often have many more sections and subsections. But the five core elements Introduction, Method, Results, Discussion and Conclusion are nearly always among them, even if called by a slightly different name. Thus what we shall present is the conventional structure of research papers in international science journals. Like any other widely accepted convention, it has the advantage that people become familiar with it and by adhering to it find that they are communicating in a common 'language'. Indeed, only by having both an international language of science, i.e. English, and an internationally accepted style/structure of communicating scientific findings, can we hope to fulfill Heisenberg's vision of *'Science as a tool for reaching understanding among peoples'*.

So let us now look in detail at the typical elements that constitiute an original experimental paper. In some papers one or more of the items mentioned may prove to be superfluous. And in theoretical papers, the experimental details and results will clearly be replaced by differently named sections, such as *'Theoretical Method'* and *'Results of Calculations'*. But the majority of the elements described below will be common to all full-length scientific papers, regardless of the discipline.

5.3.1 Title

The title should be short and to the point. It should also be informative and eye-catching. Bear in mind that readers of a journal will initially just scan the table of contents for articles of interest. It might help if you observe your own behaviour when browsing the scientific literature: At first you read only the table of contents of a journal. If a title on this list sounds interesting you will also read the abstract and perhaps the conclusions. Only if these are also interesting will you take the time to read the full paper.

A title that is too long and detailed is likely to lose the attention of many readers before they have even absorbed what the paper is about. So choose a short title, preferably no longer than one line, and mentioning only the main topic of the paper. Avoid superfluous and distracting phrases such as '*A study of ...*', or '*... measured with ...*'

Usually it is necessary to find an appropriate compromise between information and brevity. The time to choose the title is when the paper is finished; until such time, only a working title is needed.

But do be careful about the final choice of title. It may even be critical for the acceptance of the paper by the journal. A good illustration is the case of Klaus von Klitzing's paper on his discovery of the Quantum Hall Effect. The paper was submitted to Physics Review Letters. But it was rejected on the grounds that it was not interesting, did not deal with a currently hot topic, and the title was misleading. Fortunately, von Klitzing received a hint that the paper might be accepted if he were in a position to propose a better title. So he did. The new title can be seen in Fig. 5.2. At the second attempt the paper was accepted by the referee. Had von Klitzing not been convinced of the importance of his discovery – something the first referees entirely failed to recognize – the paper would not have been published, and who knows whether he would still have been awarded the Nobel Prize for this work.

5.3.2 Authors and Affiliations

After the title, the next information given is usually the names and addresses of the authors. The first author is often, but not always, the main author. In Sect. 5.1.3 we have already considered who should be named as an author and the order in which the names can/should be listed. After the names of the authors it is usual to include their affiliations, either directly or in a footnote.

5.3.3 Abstract

The abstract should be a single paragraph of 5–10 lines. Keeping it short will increase the probablity that people will read it. The abstract should summarize the contents and conclusions of the paper in a concise way. It should avoid all detailed discussion and formulas, be self-contained and should not include references.

Like the title, the abstract has the job of awakening readers' interest and making them want to read more. Thus it is important that it is formulated clearly and emphasizes the most interesting aspects of the work. In common, again, with the title, it should only be written when the paper is finished.

Since title and abstract belong together, do not test the readers' concentration by repeating the words given in the title at the beginning of the abstract. This is a common error and one to which a vigilant referee may draw attention.

5.3.4 Keywords, Classification Codes

Some, but not all, journals like to print a few keywords or classification codes after the abstract. This practice originated at the time when abstracts journals were a much used means of locating relevant papers and scanning the literature. The abstracting journals needed a means of classifying the papers they listed and, to enable this, classification schemes such as PACS (Physics and Astronomy

Classification Scheme) were invented. The PACS scheme was developed and is maintained by the American Institute of Physics (more details can be found at http://www.aip.org/pacs/) and other similar systems exist in other disciplines. Although abstracting journals are no longer widely used, you should take care to provide the correct classification and keywords so that your paper can be found in electronic databases. These days, most scientists use the comfortable approach of scanning the literature from the desk by entering the relevant keywords in a search engine or database. So think carefully about your choice of keywords. Put yourself in the shoes of another scientist for whom your paper will be of particular interest. What are the keywords that he/she is likely to type in as search criterion? By choosing the best possible keywords, you will actively contribute to your paper being read and cited.

5.3.5 The Introduction

The introduction should give a clear statement of what the paper is about. It should make clear whether the work is experimental or theoretical and should outline the aims and scope of the whole paper, why the research was initiated, what it sets out to achieve, and how it may contribute to completing a larger picture. But do not get carried away in trying to sell your research: The text of the introduction should avoid awakening expectations that are not satisfied by the paper and the results obtained.

The introduction should also briefly describe the present state of the art in the field concerned, but only to the extent needed to describe the starting point for the work reported.

Finally, in the case of longer papers with several parts, the introduction may also contain a few words about the organization of the paper.

5.3.6 Method/Experimental Details

This part of your text should provide all the information needed for the reader to understand how you obtained your results. This may involve briefly describing an apparatus or experimental procedure, whose details have already been reported elsewhere. Or, in the opposite extreme, it may report an entirely new experimental or theoretical method which itself is the main information to be communicated. In either case, it is important that the reader finds in this section sufficient information to understand, and, if necessary, check, the technique applied.

If, for example, your paper is addressing the investigation of material properties, it is vital that steps taken to prepare the sample are fully described here. Where data have been processed, or particular calculational methods employed, the reader will again want to know exactly what program was used or which approximations made. In other words, your description, together with information cited from other publications, should enable another scientist to reproduce your results.

5.3.7 Results and Discussion

This section, or sections, forms the main body of the paper. In it the relevant results should be displayed in a clear and quantitative manner, using appropriate figures and tables. Note, however, that it is not enough to display your results pictorially. Although the message contained in the plots and tables is crystal clear to you, the reader does not have the same familiarity with the topic. Thus you should also explain the salient features of the diagrams and tables in words.

Before deciding whether the results and discussion should be together or in separate sections, check the style of the journal to which you plan to send your paper. Where both options are available, then you might choose according to the nature and amount of data that you wish to present. If only one set of results is presented, then it may be useful to put the discussion into a separate section. Frequently, however, it is desirable to display several different sets of results – the one building upon the other – and to discuss them as you go along. Often a question raised in one experiment is answered in a subsequent measurement and this chronological and logical sequence will need to be reflected in the paper.

In discussing your results you should take as broad a perspective as possible, elaborating on your results and hypotheses in the light of other relevant findings and theories. If you know of previous work that is potentially relevant to yours, don't just take over the citation from another paper. Only by reading the original work can you find out whether it is really relevant; make sure that you understand it; and avoid misrepresenting it. In your discussion your results can then be compared and related to the results published by others. Other researchers will be happy to find confirmation of their own hypotheses in your paper. But what if previous work is at odds with your findings? In this case it is almost more important to mention the previous results. Discrepancies and unexplained contradictions are often a source of good new science and you should not be afraid to admit that there are still some loose ends that you cannot explain.

It is important in your presentation to distinguish clearly between the parts of the work that are new (data, hypotheses) and those that are simply reporting the state of the art. Mixing these up will make it difficult for others to correctly assign credit where it is due.

Speculations are an important and necessary part of pushing science forward. However, your discussion should rely on hard facts as far as possible. Where speculation appears to be worthwhile, keep it within plausible bounds and do not get carried away on a wave of ever more fantastic assumptions and prognoses. What is often more useful than lengthy speculation is a clear statement of open questions and some suggestions for future work that can shed further light on these questions.

5.3.8 Conclusions or Summary

After describing and discussing your results, it is usually desirable to close by drawing conclusions. Some authors content themselves with a summary of the key points at the end. Usually this contains no new information and may simply be a rewording of a few sentences from the body of the paper. Thus, like an abstract, a summary is simply a precis of the paper – with the difference, frequently, that it is written in the past rather than the present tense.

Our recommendation is to write a more detailed section entitled '*Conclusions*', in which you try to describe your contribution in a broader context. Thus, in addition to emphasizing the main results and their significance once more, the conclusion can go on to mention the wider ramifications of your work as well as open problems and possible ideas for future work. But try to keep the conclusions focused on the main question addressed in your study. Do not fall into the trap of diluting the impact of your conclusions by repeating too much of the discussion.

The conclusions, like the abstract, should be especially carefully formulated, since many readers will begin by reading only these sections. If the message delivered in the conclusions is clear and interesting, there is a better chance that readers will take the time to study your paper in detail.

5.3.9 Acknowledgements

At the end of the text, many authors take the opportunity to thank individuals and organizations who have contributed to the execution, the content or the financing of the work reported. For example, you may wish to acknowledge:

Technical support of the work by colleagues, students and technicians. All will appreciate your remembering their contribution and will be more disposed to assist with future work.

Other scientists with whom you discussed your results and their interpretation. Even though the decisive insights may have been your own, the role of colleagues as '*sounding boards*' for your ideas should not be underestimated.

Grants, awarded to you personally or to your research project, should be explicitly acknowledged giving the name of the funding source and the grant number.

If the head of department is not actually a coauthor of the paper, his or her contribution in creating an environment in which your research has flourished is also often worth an acknowledgement.

The experienced skeptic, of course, knows that one must sometimes read between the lines of the acknowledgment. '*I thank Jean Smith for expert technical assistance and Jim Williams for valuable discussions*' may also mean '*I thank Jean Smith for producing all the results and Jim Williams for telling me what they mean. But I am their boss and so this is my paper*'.

5.3.10 Appendices

If extensive calculations, supplementary data or other lengthy material is useful but not essential to the flow of the argument, it is frequently better to include this as one or more appendices. But appendices should be used with care. You should not use appendices in short communications, which need to be concise. A reference to the source of additional material is to be preferred. Furthermore, appendices should not be used to reproduce information that is readily available elsewhere, e.g. the lengthy calculational steps needed to solve a standard integral.

5.3.11 References

The reference list is an essential part of almost every scientific publication. It provides the link between your work and the edifice of already existing scientific knowledge. Without a reference list, a piece of work will appear to have arisen in a vacuum. Whilst not impossible that a new idea will have come to someone '*out of the blue*' and been tested and reported in a publication, this is certainly the exception rather than the rule.

In general, for your work to be taken seriously, it is important that you give the most important relevant references. This not only helps the reader to put your work in context, but also proves that you are well informed about the state of the art in your field. And of course referencing is a way of giving due credit to those who paved the way for your work. In the next section on formal aspects of manuscript preparation we will return to the subject of references in more detail. There you will learn, for example, how the number and choice of references should depend on the type and content of your paper.

5.3.12 Other Styles

The elements listed above provide a useful framework for structuring most full-length papers, especially those reporting on work with an experimental component. Although it is neither the structure nor the length that determines the quality of a paper, by adhering to the conventional structure, you will avoid having your paper rejected for formal reasons.

However, there are no hard and fast rules about structure. Short notes, letters and comments often have a much simpler structure. An example is the short paper, shown in Fig. 5.8, reporting the Nobel-Prize-winning work of Leo Esaki.

The papers by Shockley and Bardeen on the development of the semiconductor transistor, or by Hall on what became the Hall effect, are just as short. So clearly it is not the length but the scientific content which determines the value of a paper. What can we learn from this one-and-a-quarter-page paper that led to the Nobel Prize? The conclusion one is tempted to draw is that one merely needs to have a good idea, perform a successful experiment, write a short paper, ... and wait for the Nobel Prize.

LETTERS TO THE EDITOR

New Phenomenon in Narrow Germanium
p-n Junctions

LEO ESAKI

Tokyo Tsushin Kogyo, Limited, Shinagawa, Tokyo, Japan
(Received October 11, 1957)

the course of studying the internal field emission in very narrow germanium p-n junctions, we have found an anomalous current-voltage characteristic in the forward direction, as illustrated in Fig. 1. In this p-n junction, which was fabricated by alloying techniques, the acceptor concentration in the p-type side and the donor concentration in the n-type side are, respectively, 1.6×10^{19} cm^{-3} and approximately 10^{18} cm^{-3}. The maximum of the curve was observed at 0.035 ± 0.005 volt in every specimen. It was ascertained that the specimens were reproducibly produced and showed a general behavior relatively independent of temperature. In the range over 0.3 volt in the forward direction, the current-voltage curve could be fitted almost quantitatively by the well-known relation: $I = I_s[\exp(qV/kT)-1]$. This junction diode is more conductive in the reverse direction than in the forward direction. In this respect it agrees with the rectification direction predicted by Wilson, Frenkel, and Joffe, and Nordheim 25 years ago.[1]

The energy diagram of Fig. 2 is proposed for the case in which no voltage is applied to the junction, though the band scheme may be, at best, a poor approximation for such a narrow junction. (The remarkably large values observed in the capacity measurement indicated that the junction width is approximately 150 angstroms, which results in a built-in field as large as 5×10^5 volts/cm.)[3] In the reverse direction and even in the forward direction for low voltage, the current might be carried only by internal field emission and the possibility of an avalanche might be completely excluded because the breakdown occurs at much less than the threshold voltage for electron-hole pair production.[4] Owing to the large density of electrons and holes, their distribution should become degenerate; the Fermi level in the p-type side will be 0.06 ev below the top of the valence band, E_v, and that in the n-type side will lie above the bottom of the conduction band, E_c. At zero bias, the field emission current $I_{v \to c}$ from the valence band to the empty state of the conduction band and the current

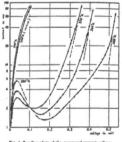

FIG. 1. Semilog plots of the measured current-voltage characteristic at 200°K, 300°K, and 350°K.

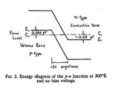

FIG. 2. Energy diagram of the p-n junction at 300°K and no bias voltage.

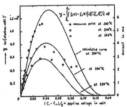

FIG. 3. Comparison of the current-voltage curves calculated with the measured points at 200°K, 300°K, and 350°K.

$I_{c \to v}$ from the conduction band to the empty state of the valence band should be detail-balanced. Expressions for $I_{v \to c}$ and $I_{c \to v}$ might be formulated as follows:

$$I_{v \to c} = A \int_{E_c}^{E_v} f_v(E)\rho_v(E)Z_{v \to c}\{1 - f_c(E)\}\rho_c(E)dE,$$

$$I_{c \to v} = A \int_{E_c}^{E_v} f_c(E)\rho_c(E)Z_{c \to v}\{1 - f_v(E)\}\rho_v(E)dE,$$

where $Z_{v \to c}$ and $Z_{c \to v}$ are the probabilities of penetrating the gap (these could be assumed to be approximately equal); $f_c(E)$ and $f_v(E)$ are the Fermi-Dirac distribution functions, namely, the probabilities that a quantum state is occupied in the conduction and valence bands, respectively; $\rho_c(E)$ and $\rho_v(E)$ are the energy level densities in the conduction and valence bands, respectively.

When the junction is slightly biased positively and negatively, the observed current I will be given by

$$I = I_{v \to c} - I_{c \to v} = A \int_{E_c}^{E_v} \{f_v(E) - f_c(E)\}Z\rho_c(E)\rho_v(E)dE.$$

From this equation, if Z may be considered to be almost constant in the small voltage range involved, we could calculate fairly well the current-voltage curve at a certain temperature, indicating the dynatron-type characteristic in the forward direction, as shown in Fig. 3.

Further experimental results and discussion will be published at a later time. The author wishes to thank Miss Y. Kurose for assistance in the experiment and the calculations.

[1] A. H. Wilson, Proc. Roy. Soc. (London) A134, 487 (1932); J. Frenkel and A. Joffe, Physik. Z. Sowjetunion 1, 60 (1932); L. Nordheim, Z. Physik 75, 434 (1932).
[3] McAfee, Ryder, Shockley, and Sparks, Phys. Rev. 83, 650 (1951); C. Zener, Proc. Roy. Soc. (London) 145, 523 (1934).
[4] S. L. Miller, Phys. Rev. 99, 1234 (1955); A. G. Chynoweth and K. G. McKay, Phys. Rev. 106, 418 (1957).

Fig. 5.8. The entire paper in which Leo Esaki reported the work for which he was later awarded the Nobel Prize [Phys. Rev. **109**, 603 (1958)]

5.4 Formal Aspects of Manuscript Preparation

A scientific paper involves a lot more than just writing text. In this section we deal with the various formal aspects that pertain to the preparation of a journal article.

5.4.1 Figures and Tables

It goes without saying that the figures should be clearly drawn and labelled and that all symbols and writing should be easily legible. For aesthetic reasons it is also desirable for all figures to be produced in the same style using the same fonts, line thicknesses and styles of symbols, shading, etc. This is facilitated by using the same drawing program for all figures. Even if such uniformity of appearance is not possible, pay attention to the aesthetic impression made by your figures. Are they pleasing to behold? Beware, especially, of using very thin lines. Although they may appear clearly on a computer monitor, lines of thicknesses below 0.2 mm may not be properly reproduced in print. On the other hand, very thick lines and very large or bold fonts are ugly and distract the reader from the intended information content.

As already mentioned in Sect. 5.1.5, it is often sensible to choose one key figure for your paper. This should be prepared with the greatest care. If your results are really significant, the figure summarizing them will be reproduced in other publications and still recognized by future generations of scientists.

For results presented in graphical form it is recommended that the figure should have a frame on all four sides, with the units and scale marked on all axes, as shown in Fig. 5.9. A widely accepted notation for axis labelling is to write the name of the quantity in full, followed by its units in round brackets, e.g. frequency (Hz). This form is also well suited as a column header in a table. For good legibility the writing on the figures should be in the range 9-12 point. To avoid confusion, all symbols appearing on figures should be identical in style to those used in the text.

Every figure and table should be given a concise but informative caption, which explains all symbols and abbreviations particular to the figure/table concerned. All figures and tables should be referred to in the text. In a camera-ready manuscript you should try to position the figure as close as possible to its discussion in the text, but preferably not before this and certainly not in a preceding section.

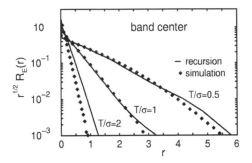

Fig. 5.9. Example of an almost ideal figure

5.4.2 Units

For most papers in the natural sciences and engineering, the appropriate units are those of the SI system. But there are some exceptions to this rule and it is advisable to use whichever system of units is most widely accepted in your field. Beware of using an outdated or minority system of units, since this will make your paper less accessible and may cause delays if revisions are requested as a result. The basic units of the internationally established Système International (SI) are s, kg, m, A, K, Cd, and Mol. A set of derived units, which is approved by the International Union of Pure and Applied Physics (IUPAP), includes Hz, N, V, W, Ω, J, F, and Wb. Note that in camera-ready manuscripts (including electronic files that will be used for reproduction), it is conventional to set units in upright (roman) characters, not italic.

5.4.3 Abbreviations

Abbreviations – including acronyms – are a very useful tool in science, both for keeping a text to a manageable length and also for reasons of clarity. However, they must be used with care. Authors who are familiar with the jargon and abbreviations used in their field are often inclined to forget that novices and outsiders will not recognize many of these. Thus, rather than improving the clarity of a paper, overuse of abbreviations and failure to define them, can mean that potentially interested readers will give up frustrated. If an abbreviation is internationally accepted and known to, say, every PhD student in the discipline (e.g. dc, DNA, ppm) then it is permissible to use it without explanation. Every other abbreviation or acronym should be clearly (and correctly!) defined at the point of the main text where it is first used. It is usual to write the expression out in full and then introduce the abbreviation in brackets afterwards. For example: '*Reflection high energy electron diffraction* (RHEED) *is an established technique for studying …*' Thereafter, one can use the abbreviation on its own.

In books, and even in some longer review papers it is useful to include a separate list of abbreviations.

5.4.4 Symbols

> '*The symbols are so illuminating that the fact that the text is incomprehensible doesn't much matter.*'
>
> A.N. Prior

Like abbreviations and acronyms, symbols should be defined where they are first used unless they are absolutely standard and will be known to all readers of your paper. Use well-established symbols wherever possible, although you may occasionally have to deviate from convention in order to avoid ambiguity. Sometimes it is impossible to use a unique symbol for every quantity. Typically, the 'e' for exponent is also used to represent the charge of an electron. This causes no confusion, however, since the meaning is invariably clear from context (and usually the former is set roman and the latter italic). However, for more unusual quantities, and wherever there is a danger of confusion, the symbols must be kept distinct. A good way is often to use subscripts to identify the different usages.

5.4.5 Subject Index

A subject index is needed only in books and book chapters. With modern word processing software, the subject index can be generated automatically if you insert the entries in the text as you go along. Software such as LaTeX is especially powerful in this respect.

5.4.6 References

Importance of References. The reference list is an essential part of any scientific paper. It serves several purposes, of which the following are the main ones:

▷ It allows the reader to find additional information relevant to the work being reported, but not essential to the presentation. This typically includes papers and other literature covering general background to the problem, details of the experimental method, data and other findings upon which your analysis is based, previous interpretations (confirming and/or conflicting), derivation of equations, and any other work that is directly relevant;

▷ It gives credit to those scientists who have previously contributed to the field and whose work you are building upon; typically the seminal papers by the founders of a particular field are cited in the introduction;

▷ It demonstrates that you are familiar with the relevant literature and thus with international efforts in your field. An appropriate reference list will give readers more confidence in your paper.

How Many References Are Needed? The number of references appropriate to a publication will depend strongly on the field concerned, the breadth of the work reported and the type of article used to communicate it.

For papers in physics, as a rough guideline, we suggest that 20 references is appropriate for a letter or short note, 50 for a full length original paper, and as many as 200 or more for a review paper attempting to comprehensively survey a whole field. For publications in biology or medicine the average number of references is usually about double the figure for physics. In chemistry, however, authors tend to cite only about half as many references, and in mathematics fewer still. The reason, probably, has to do with the overall size of the literature in these fields and the 'fineness' of the respective mesh of knowledge.

Collecting References. Producing an appropriate reference list is not difficult if you are well abreast of the relevant literature, a desirable way to be – also for the sake of your own scientific knowledge and planning (see below).

To ensure that you have the appropriate papers and citations to hand, not only for preparing reference lists, it is useful to keep a list of useful papers with comments about their content and copies of the most relevant ones. But you should collect these papers in an organized fashion, the hard copies in folders for different topics and the electronic versions in corresponding directories on your computer. A little time invested in setting up this resource is time well spent as it can save you hours of searching in vain. Searching, for example, for that recent paper (or was it last year?) by 'Wotsisname' and 'Thing' in that new journal 'Nanosomething'...

Since many papers are now available online, these, or at least their abstracts, can be stored as links. As your research progresses, you will find that the number of relevant papers multiplies rapidly. Hence a well-organized filing system, whether physical or electronic, becomes indispensable.

Keeping up with the Literature. Rather than having to begin an extensive literature search when you write your paper, it is far better to keep yourself continuously informed by scanning the main specialist journals at regular intervals, preferably every week. If you plan one or two hours a week, this should give you enough time to at least skim through the important new papers in your field. Again this is time well invested, even though it may seem to be slowing you down when you have other urgent work. These days one can scan the literature far more efficiently than 20 years ago. Increasingly, you can make use of virtual journals and alerting services to find out what new publications have appeared in your field; see Sects. 6.2.3 and 6.3.2.

Secondary Citations. It is wise to have read all the papers that you cite in your own publications. Relying on secondary sources can lead to errors in the citation, and – worse still – even to misrepresentation of the content. If, for example, you reproduce the printing error in a reference taken from another journal, it will become obvious that you simply lifted the reference without without reading the paper. As a general rule, if a paper is worth citing, it is also worth reading. If you cannot access a paper because your library does not subscribe to the journal, you can try to order it from another library or ask the author to supply it by e-mail. In the 1980s and earlier it was common practice to ask authors to send offprints of their papers by post. Most recent papers can now be acquired from the publisher electronically. For those institutes that do not subscribe to the journal, individual papers can be viewed and/or downloaded via pay-per-view. Don't miss a good paper which might have far-reaching consequences for your work just because it isn't on the shelf.

Format of Reference List. The formatting of a reference list for a publication is an art unto itself. There are several different styles of referencing, the main ones being (a) sequential numbering in order of citation and (b) citation by name and date with alphabetical listing. Similarly there are different styles of indicating the citation in the text (superscript, name in round brackets, number in square brackets ...). Which of these systems you choose, will depend on the journal to which you want to send your paper. If you have a lot of references and they are going to be numbered, there is an obvious danger that a reference added as an afterthought at the beginning can cause a good deal of work renumbering all subsequent references. For this reason, it is often best to introduce the numbering only when you are finished, or to use a program such as LaTex, which can take care of the numbering automatically.

On the reference list itself, there is an even greater variety of formats possible. Each journal and each publisher seem to have their own preference. So take a minute to check the style used in the journal you have chosen – it may speed the passage of your paper through the editorial department if you can adhere to its rules.

For numbered references, a fairly standard format for listing a journal article is:

1. J. Bardeen and W.H. Brattain, Phys. Rev. 74, 230 (1948)

And the foregoing text citation will appear, typically as [1] or [1].

While varying in details of punctuation, almost all numbered reference styles will have the following features in common:

▷ authors' names given by initials preceding surnames

▷ name of the journal abbreviated; lists of standard abbreviations on the Internet; a fairly exhaustive one, due to the CalTech: http://library.caltech.edu/admin/abbreviations/

▷ volume number of the journal printed in bold

▷ first page of the article – and occasionally the last – specified

▷ year of publication indicated in round brackets.

In the name-date system, on the other hand, the authors are listed alphabetically and the style is typically:

Bardeen, J., Brattain, W. H. (1948): Phys. Rev. **74**, 230

with the surnames in front of the initials and the date immediately following the author names. The text citation is then (Bardeen and Brattain, 1948).

Different styles are required when referring to books, articles from books, PhD theses. For a book, one commonly encounters, in a numbered system:

1. P. Yu, M. Cardona, *Fundamentals of Semiconductors*, 3rd edn., (Springer, Berlin, Heidelberg 2001), p. 377

and in the name-date system:

Yu, P., Cardona, M. (2001): *Fundamentals of Semiconductors*, 3rd edn., (Springer, Berlin, Heidelberg), p. 377

Provided you have given all the information needed for a reader to find the source referred to, it is unlikely that your paper will be returned for revision of the reference list. But it is certainly useful, also for the reader, if you stick to a consistent style and one as close as possible to that of the journal to which you are submitting the paper.

Other details about referencing style and also use of *et al.* can be found in the Chicago Manual of Style.* Most publishers also issue their own guidelines, either in the journals themselves, or on their web page.

Note that publications that will also appear online are making increasing use of the '*Crossref*' facility (a system allowing the user to link directly from a citation to the abstract of the publication itself – see Sect. 6.3.2). Where this is to be implemented, there are special rules about the format of the references.

5.4.7 Camera-Ready Manuscripts

In academic publishing, especially of scientific and technical subjects, most publishers nowadays request the submission of camera-ready manuscripts for both books and journal articles. The motivation for this is to achieve quicker publication at more moderate cost. Increasingly, the same electronic data that are submitted for printing are used to generate an online version of the paper or book.

To help authors achieve the desired layout and typography for their texts, most publishers provide LaTeX macros and/or Microsoft WORD templates. These enable you to write your text directly in the required format. The better and more powerful typesetting program is undoubtedly LaTeX, but, being more sophisticated, it also takes some time to learn. However, if you are planning to write papers and other scientific texts in any quantity and over several years, it is almost certainly worth taking the trouble to learn LaTeX.**

The major advantages of this program are (1) that it can automatically take care of all the numbering (sections, equations, figures, references, etc.); (2) it yields top quality typography, and is especially powerful for typesetting equations and mathematical

---

*   The Chicago Manual of Style: The Essential Guide for Writers, Editors and Publishers, 14th Edition (University of Chicago Press, Chicago 1993)

** See, for example, Helmut Kopka, Patrick W. Daly: A Guide to LaTex, 4th Edition, (Pearson, 2003)

Fig. 5.10. Simultaneous writing and typesetting is becoming the norm

symbols; (3) within the mathematics and physical-science communities it is well established and widely used by most institutes and by publishers. When you have mastered LaTeX, you will find that it is a joy to use!

5.4.8 Some Notes on Copyright

The concept of copyright originated with the advent of printing in the late fifteenth and sixteenth centuries. It was first formalized in England in 1710 when the British Parliament enacted the Statute of Anne. Designed mainly to protect the rights of printers and booksellers, this statute also established the principle of authors' ownership of copyright.

Copyright Law Today

Possession of copyright gives the owner the right to reproduce, distribute, perform, display or licence the work concerned. Conversely, the copyright law makes it an offence for other persons or organizations to exploit the work in these ways. In most countries, copyright protection endures until 70 years after the death of the longest surviving author. (Exceptions are Japan and Canada, who adhere to the minimum period of 50 years that was specified in the Berne Convention of 1971.)

Although originally intended only to protect the writings of authors, changing technology has meant that the word '*writings*' has now come to take on an ever wider meaning. Copyright law now covers architectural design, software, graphic arts, motion pictures and sound recordings. To be eligible for copyright a work must be original and in a concrete medium of expression, i.e. fixed in a tangible or material form.

The law on copyright is very complex and, since the birth of the Internet is being constantly stretched to provide the necessary interpretations and guidelines for the protection and transmission of electronic multimedia material. Currently the World Intellectual Property Organization (WIPO) is addressing a number of proposals for changes to meet the global information infrastructure.

Copyright in Science Publishing

As far as regular scientific papers and books are concerned, the copyright is generally transferred by the author to the publisher, who then acquires the right to print and distribute the work. In return for transferring the copyright to their work, authors benefit from the service of having their work printed and disseminated.

However, when you are preparing to submit a paper for publication, the main copyright question that you will be faced with relates to the use of previously published material, in particular figures. If you plan to use diagrams or other figures, even ones of your own, that have already been published elsewhere, you should ask the copyright-holder (usually the original publisher) for permission. This is almost invariably granted on the condition that you give a full reference to the source. This means, at the very least, that you should write '*From* [XY]' at the end of your figure caption, and give the full reference on the reference list.

Occasionally, publishers will ask for a fee for reproducing an image. This might be the case, for example, if you wish to reproduce an attractive four-colour picture of a galaxy on the front of your book. But for a figure showing a set of data-points on a graph, permission is usually given free of charge.

For the convenience of their authors, many journal and book publishers provide a standard letter for requesting permission to reproduce copyrighted material.

What about figures that have been modified? This is a rather grey area. If you have added new elements to, redrawn and/or re-labled a figure that was published in a previous paper, then – pro-vided the changes are significant – it is not necessary to request permission, since you have created a new figure from the same information (note, the information itself is not subject to copyright, only the creative process of selecting, organizing and displaying it!). In this case the standard practice is to acknowledge the source of the original figure by writing '*After* [XY]' at the end of your figure caption and, again, giving the full reference on the list.

5.5 Submission, Refereeing, Revisions

Submission
Once your paper has been written, usually with a particular journal in mind, and all authors are satisfied with its content, the time has come to submit it for publication. Different journals have different rules about how many copies of the paper are to be sent to whom. Usually, it is now also possible to submit papers in purely electronic form.

Refereeing
The publisher will usually acknowledge receipt of your paper immediately and then pass it on to a member of the scientific editorial board for refereeing. The scientific editors will referee some papers themselves; others, which lie somewhat outside their speciality, will be passed on to one or more independent referees.

Can you influence the choice of referees for your paper? Although you could, conceivably, make suggestions for potential referees when you submit your paper, this practice is frowned upon by editors and generally these people will not be asked. But if you have had negative experience with one or more scientists, perhaps competitors, then – if you must – you can request that your paper is not sent to them to referee. Most journals will accept such a request.

It is worth remembering that journal editors are often hard-pushed to find appropriate referees for specialist research topics. So if your work falls into this category – and most does – then you can perhaps help the editor. One of the main ways in which referees are chosen is to check the references of a paper and see who is cited as having done prior and/or related research on a similar topic. By citing particular papers in an appropriate context you might just be lucky and get the referee of your choice.

Regardless of how the referees are chosen, the process of peer review is essential for maintaining standards, not only of individual journals, but throughout science. It is a tried and tested philosophy and one of the pillars of modern science. But like all systems it has its imperfections too. In our experience, one of the main problems encountered with peer review is the delay that it can cause and the corresponding frustration for authors.

Although referees are always asked to respond rapidly, it is not unusual for several weeks and a couple of reminders to be required before an answer is received. If you have been active as a referee yourself, then you will know that careful refereeing is a time-consuming task that often has to be squeezed in among many other scheduled duties. In this respect too, some journals are better than others. The more prestigious ones are, unsurprisingly, more successful at attracting good and reliable referees.

But sooner or later one or more referee reports will be forthcoming.

Acceptance or Rejection

Based on the referee reports, the journal editor will suggest one of three possible courses of action:

- ▷ the paper, in its present form, is rejected
- ▷ the paper is accepted subject to certain revisions
- ▷ the paper is accepted as it is.

The rate of rejection is naturally highest for the most prestigious journals. Nature and Science reject about 10 papers for every one that they print. If the scientific content is incorrect or dubious then the paper will (hopefully) be rejected by every self-respecting peer-reviewed journal. More prestigious journals can afford to be more demanding than this: In addition to being scientifically correct, the topic has to be appropriate to the readership of the journal and the paper must contain sufficient new and interesting material to be worthy of publication. In this way, the more prestigious journals manage to fill their pages with articles of a consistently good quality.

A fairly large proportion of papers are returned for revisions of one kind or another. This can lead to long delays in publication and, as an author, you can reduce the likelihood of such delays by making sure that you observe the formal conditions for publication. For example, papers are quite often returned because they are too long for the journal in question; because the title and abstract make promises that are not fulfilled by the work reported; because essential references have been omitted (if these 'essential' references are all by the same person, you might draw some conclusions about who wrote the referee report!); or because the quality of the English is too poor. By checking all these aspects prior to submission, you can often save a good deal of time.

Disputes with Referees

If you become embroiled in a long and unsatisfactory exchange with one or more referees who don't like your style of doing things, or who appear to have missed the point of your work, it is sometimes worth asking the journal to engage an alternative referee. But even when you disagree vehemently with unkind criticism from a referee, keep your cool! Answering in a similarly aggressive tone will do nothing to convince the referee and will also make the publisher and/or responsible journal editor less sympathetic to your plight.

Sometimes there is no alternative but to withdraw your paper from consideration and send it to a different journal. For obvious reasons it is not good practice to submit a paper to more than one journal at once.

But if you receive the following referee report, there is probably no sense in bothering to resubmit your paper: *'This paper contains much that is new and much that is true. Unfortuantely, that which is true is not new and that which is new is not true.'*

A Happy End

If all goes well and your paper is accepted as is – congratulations, you have properly followed the advice given in these pages!

Once your paper has been accepted, it may still take another 2–4 months before you see it in print. Some journals have a *'rapid communications'* section which is used to accelerate the publication of the most important short papers. If you attach importance to a quick publication and believe that your paper is of sufficient merit, then you can explicitly submit it as a *'rapid communication'*. With the advent of publication on the internet, many journals now publish articles *'online first'*. In such cases the turnaround time can be reduced to a matter of weeks, or even days, once a paper has been finally accepted.

5.6 Writing for Profit?

Most scientists regard publishing their results as a normal part of their job. Since they recognize that good publications play an important role in establishing their reputation and furthering their careers, they usually expect no additional payment for preparing papers and often work on their manuscripts in their leisure time.

In some cases, e.g. review manuscripts, book chapters and whole books, authors are rewarded with a small flat fee or sales-oriented royalty. However, with a few exceptions, this income does not make scientific writing a lucrative activity. Only the authors of standard textbooks and successful popular science books really make much money from their writing. A book editor known to us, admittedly a very meticulous one, once calculated that the flat fee

Fig. 5.11. Not everything printed on paper is of equal value

he received for his work provided him with an hourly income of 5 US cents! (Alas, not all editors are so careful.)

So for most scientists the real profit to be made from writing is a non-monetary one: namely, the satisfaction of having contributed in a small way to documented human knowledge, and of course the pleasure and pride of gaining a good scientific reputation. To profit the most from your efforts means that you need to 'sell' your work for the best return, i.e. try to have it published in a high-ranking and high-impact journal. Aim high, but be realistic. If you are not in a hurry, you can always try another journal later. Remember, the value of the printed word is no more than the value of the paper, unless those words are read by someone to whom they mean something.

5.7 Dos and Don'ts

Much of the advice given in the aforegoing sections on preparation of scientific papers can be summarized by the following list of dos and don'ts:

DO: Choose the proper content for your paper.
DON'T: Overload it with irrelevant and peripheral information.

DO: Choose an appropriate type of publication.
DON'T: Sacrifice the chance of a top quality original paper by rushing to get a short note into print.

DO: Be careful about the time of publication.
DON'T: Reduce the impact of your results by publishing them too early or too late; and don't miss the opportunity of a patent application.

DO: Aim to publish in a highly reputed international journal.
DON'T: Hide important results in a low-level, limited circulation, or non-English language journal.

DO: Publish true and confirmed results only.
DON'T: Sacrifice accuracy for the sake of speed.

DO: Include as authors those who made significant contributions to the work reported.
DON'T: Forget to acknowledge further sources of scientific, financial and moral support.

DO: Observe the ethics of intellectual property and the legal rules of copyright.
DON'T: Make yourself liable to any charge of plagiarism.

DO: Structure your paper in a logical and accepted fashion.
DON'T: Discuss your results before presenting them.

DO: Discuss your results in the light of other related work.
DON'T: Forget to give appropriate references.

DO: Choose an appropriate and 'punchy' title and make sure that abstract and conclusion are informative and concise.
DON'T: Try to squeeze too many details into title, abstract and conclusion;

DO: Pay attention to the formal requirements of the journal, e.g. length, formatting, language.
DON'T: Be disappointed if your paper is returned for revisions.

6 Electronic Publishing

The widespread use of personal computers and the internet has given birth to a new form of publishing, commonly referred to as electronic publishing. In the same way that e-mails are replacing letters and telephone calls as a convenient means to communicate quickly, electronic publishing is becoming a vital component in the fast publication and dissemination of scientific information.

Of course, electronic publishing is not limited to science, but it is the scientific publishers and scientists who have taken the lead in developing this new type of publishing. It is for them, too, that it brings the greatest benefits, since science is a fast-moving endeavour and scientists are always keen to publish and read the latest results with a minimum of delay. Thus, one might claim that electronic publishing will also accelerate the progress of science itself.

There is much debate in the scientific publishing community as to whether electronic publishing will one day mean the complete disappearance of paper journals and books. At present the consensus is that books and journals will remain a valuable means of transmitting and archiving information, but that electronic products will become the most important source for those wishing to quickly find or communicate new results. The arrival of man did not mean the extinction of the apes; each had, and still has, its own role to play on the stage of nature.

So too can one imagine the coexistence of electronic and conventional publishing forms. As we will see later, many scientific information products are published simultaneously in paper and electronic form. This gives the user a convenient choice. For quickly checking a reference or confirming some details, it is convenient to display an article on your computer screen and be able to search it for keywords. However, if you are studying for an exam, doing a literature review, or learning about a new field, it may suit you better to take away the printed journal/book and read it at your leisure: in bed, in the bath, on the beach, … places hard for a computer to reach. Thus, for journals at least, parallel publication of paper and electronic versions is currently the most common policy of the big publishers.

6.1 Milestones in the Development of Electronic Publishing

Development of Electronic Publishing

The birth of electronic publications can be traced back to about 1985, from which time onwards there has been steady progress:

Fig. 6.1. How times are changing (based on Raffael's School of Athens)

~1985:   first books with CD-ROM appeared

1990:   servers for the immediate announcement of scientific re-
         sults are installed, e.g. the Los Alamos preprint server
         (Ginsparg server at Cornell University: http://arxiv.org)

1990:   publishers provide the first e-mail alerting services, draw-
         ing customers' attention to articles of interest in printed
         journals

1995:   online versions of scientific journals become available,
         additionally to the print version and also electronic-only
         journals. This date also marks the point at which scientif-
         ic publishers began major investment in the development
         of electronic publishing

1996:   attempts made to launch online-only journals

1998:   books start to appear as both printed and online product.
         E.g. Springer's series Topics in Applied Physics

1999:   virtual journals established (Sect. 6.2.2)

1999:   cross-linking of journal articles begins

2000:   first online-only books

2001:   international agreement reached by a group of about 500
         leading scientific publishers, allowing them to cross-refer-
         ence each other's articles

2002:   development of 'living books'

2004:   move towards Open Access gains momentum (Sect. 6.7)

2006:   fully electronic journals and books with a paper-like dis-
         play expected to become available(Sect. 6.6.2).

Breakthroughs for Electronic Publishing. None of these developments would have been possible without accompanying technological advances. One can identify a number of breakthroughs that
(with hindsight) have proved to be prerequisites for the successful
introduction of electronic publications. The most important were:

▷ ASCII code

▷ personal computer

▷ increasingly powerful microchips

▷ WYSIWIG: what you see is what you get

▷ improved data transmission: networks, telecom, wireless

▷ multimedia PCs

▷ WWW

▷ introduction of standards: HTML (hypertext markup language), XML (newer: extended markup language), DOI (digital object identifier)

▷ electronic paper; at the time of writing, still at the prototype stage.

6.2 Today's Electronic Media and Products

It is useful to distinguish between media and products, although to some extent these overlap with one another. Media are essentially the carriers of the data, but the term is also used to describe the type of data (audio, visual, ...), whereas the products are what define the content, organization, updating modus, etc. of the information. Here are some of the most important examples in each category:

6.2.1 Media

Webserver on which data for books/journals can be stored for viewing and/or downloading via the internet

Multimedia PC allowing not only images but also videos and sound to be part of the information received

CD ROMs and DVDs sold individually and used to store information (software, music, films …) on user's PC

Electronic Paper. This revolutionary new technology has not yet become available commercially. However, due to its immense potential importance, we will devote a section to it later on (Sect. 6.6.2).

6.2.2 Products

The introduction of electronic publishing and new media has not only changed the way in which books and journals are made available; it has also given rise to complete new product categories, such as living books and electronically available topic-oriented review journals. The main product types currently available are the following:

Offline or Stand-Alone Products

CD ROMs (capacity up to 800 MB) whose contents can be a database, software, book-like text (often crosslinked to allow the reader to jump to points of interest), music CDs. In the early years, CD-ROMs were often sold in conjunction with a book as a means of increasing the amount of data available or the functionality of the combined product. For example, computer programs for running simulations could be provided with a scientific textbook.

DVDs (capacity 4.7 to 17 GB) are similar to CD-ROMs, but due to their higher capacity also allow to store complete films with bonus material. They play no significant role in scientific publishing at present.

6.2.3 Online Products

Electronic Journals. Parallel electronic versions of printed journals were one of the first types of online publication. They are still one of the most important eletronic media and are generally made available to all those who subscribe to the printed version. A pay-per-view option is sometimes included to allow limited access to non-subscribers.

Online-First Journals. It is now a frequent practice for the electronic version of a scientific journal to be published slightly or sometimes significantly in advance of the printed version (online-first publication). As publishers gained more experience in preparing material for electronic access, it became possible to make the electronic version of many journals available a few weeks in advance of the printed version (simply because the time needed for completing a volume, printing, binding, and distribution is saved). This is clearly a useful service for scientists hungry to read about the latest developments in their field. If you are very eager to have your paper accessible to many scientists as soon as possible, choose a journal which provides online-first publishing.

For publication in the printed version, only the final page numbers, citation line and online publication date are added. However, this is not a preprint service. When the publication is put into the net it is in the final form. Changes of contents and withdrawal of an article is then impossible. Online-first articles are equivalent to printed articles. Online-first articles are already citable via DOI (see Sect. 6.3.2). The DOI is also used to create hyperlinks to other online-first articles.

Online-Only Journals. A number of publishers and academic institutes have founded journals in which the electronic texts have no corresponding printed counterpart. These few journals are online only and the information is only archived electronically.

After a few years of experience with such products, it appears that many scientists are still rather reluctant to publish in such

Fig. 6.2. Online-only journal: *EPJ direct* (EPJ: European Physical Journal)

journals. Perhaps understandably: Possible changes in data formats and new hardware could mean that today's electronic only products become inaccessible to future generations. With no guarantee in this respect, it is clear that parallel publication of print and online versions is favored. Indeed, some online-only journals have now reverted to parallel publication.

Download Products. Some publications in the above three categories are also offered as a special download product. The only difference, typically, is that the download version has a higher resolution so that images (requiring only a low resolution for monitor-viewing) can now be printed in the higher quality that one expects of books and journals.

Virtual Journals. A virtual journal is simply a web page that serves as a multijournal guide to papers appearing in a particular field. A good example is the virtual journal '*Nanoscale Science and Technology*' which provides links to abstracts from relevant papers from different journals. At present 53 journals from 14 publishers participate. Every year more and more virtual journals in other fields of research are founded.

This service is free. Register for the virtual journal in your field of research – if there is already one – to make sure not to miss important papers.

Living Books. These are online books covering active and rapidly changing fields of research. They can be continually updated and extended by the authors. Every year or so, or whenever the need arises, the latest version is printed and made available as a conventional book. But readers of the electronic version can stay up to date without waiting for the next printed edition.

'Jooks' (or 'Bournals'). As the name suggests this is a product that is a hybrid between a book and a journal. Like many books, the jook is a collection of chapters, usually by different authors, detailing various aspects of a central topic. But the collection can grow and its contents are flexible. As new knowledge becomes available the chapters can be continually updated. Major new breakthroughs may warrant the addition of new chapters. Older chapters whose content become obsolete can be removed. Every now and then – say every two years – the publisher may decide to print the collection as it stands and offer it to those who wish to have the information in conventional book form.

6.3 Use and Benefits of Online Publications

6.3.1 Prerequisites for Use

If you or your institute wish to take advantage of online publications you need to possess some basic infrastructure and make the appropriate purchase agreements with the publishers.

Equipment. The first prerequisite is the availability of modern PCs with the necessary software. To fully exploit the possibilities offered by the online data, and to optimize your own electronic publications, a range of programs are needed, which should ideally allow all the following formats to be produced and viewed:

▹ pdf: portable document format

▹ HTML: hypertext mark-up language

▹ TeX, LaTeX, IBM Techexplorer

▹ images: gif, jpeg, png, tif ...

▹ sound: mp3, wav, ...

▹ movies: Apple's QuickTime, mov, mpeg, ...

> chemical structure: pdb, yxz, mol

> virtual reality: wrl, vrml with viewer (Live3D, e.g.)

Even an isolated PC can exploit all these programs to give access to images and sounds stored on a hard disk or CD. But to access online data, the PC has to possess the internet protocol (TCP/IP) and be connected by telephone (directly or indirectly via an intranet) to the internet. This second prerequisite, however, is fulfilled by virtually every PC these days: New PCs are delivered with the necessary programs (internet protocol and browsers) pre-installed.

Costs for the User. Finally, the electronic products offered by scientific and most other commercial publishers are rarely free of charge (but see also Sect. 6.7 on Open Access). Some allow free access to all or parts of the text; others are only available to subscribers.

Although tables of contents and abstracts are available to everyone, to access the full information a licence must be bought from the publisher. Once this is issued, users can login to the site concerned by giving their user name. A password provided by the publisher enables the PC concerned, or in the case of a multi-user licence, a group of PCs, to access and display the required data.

Another way to get the full text of an article is to pay per view if you do not have access to the journal.

Now you are all connected up and ready to go. What benefits can you expect?

Fig. 6.3. http://www.firstgate.com: click and buy. One of several methods for purchasing electronic journal articles

6.3.2 Advantages of Online Publications

For the user, electronic publications offer several advantages over conventional scientific publications in books and journals on paper. These are summarized below.

Convenience. To find the information you are seeking, you need not move from your desk. Indeed, if an institute subscribes to the journal, access is usually possible from any one of the PCs within the building.

Supplementary Material. Because there is virtually no limit on the amount of data that can be stored and transferred, online publications are not restricted to text and black and white diagrams. Full four-colour pictures, internet links to other sources, video clips, animations, and even sound can all be used to extend and enrich the content of an online publication – at no additional cost.

Speed. Online publishing can be exploited to make information available significantly faster than is possible with conventional paper products.

Preprint servers can be used to gain access even before articles are accepted for publication.

An advantage of electronic-only and online-first journals is that the papers become available more rapidly and, in principle, more widely. Even including the refereeing process, the electronic ver-

Fig. 6.4. Instant access to scholarly information

sion of an online-first journal article is typically available some weeks before its paper counterpart. This is because the time needed for printing, binding and delivering the paper copy (often several weeks in total) is saved.

Easy Searching. Using the search facilities of the publisher's website, or indeed of an internet search machine, it is possible to locate articles that are relevant to the topic of interest. This is much simpler than going to the library and scanning the contents and indexes of the paper journals or even of compilations such as abstracts journals. Using classification schemes such as PACS (Physics and Astronomy Classification Scheme) electronic articles can be accurately classified and retrieved. The use of such classification schemes becomes particularly important for electronic publications and – on the internet at least – is becoming the main tool for serving scientists with the information of interest.

By searching the internet for particular keywords, there is an excellent chance that you will find the relevant papers of interest. With conventional journals, this requires far more perseverance – the study of numerous abstracts and indexes – and, indeed, is largely a matter of luck.

Free Information. Even though your library may not subscribe to a particular journal either in printed or electronic form, the electronic version usually gives free access to the table of contents and abstracts of all articles published. Thus you need never miss an important paper. Pay-per-view arrangements can be used to access the full text of individual articles of particular interest. Similarly, for electronic book publications, you can usually access the preface, table of contents, and introduction free of charge. Sometimes it is also possible to access individual chapters on a pay-per-view basis.

Alerting Services. By making use of an alerting service you can even avoid the need to search for articles on your special subject.

You can register for free alerting services at the webpages of most international scientific journals. You can fine-tune your information preference and narrow down your selection to certain specialist areas. In some cases you can also specify the frequency of e-mails sent to alert you to new publications. When sensibly used, alerting services can offer a great improvement in efficiency when it comes to keeping up to date on research in your area. We recommend signing up for the alerting services of all the journals which are of scientific interest to you, including journals not available in your institute. Then you can be sure not to miss any important developments in your research field.

Personal Archiving. For journals published in electronic form, it is a trivial matter to keep track of papers of interest. You can store either the link to the paper or the pdf file of the paper itself on your computer. But do it in an organised way to find it easily when you are looking for this information later.

For libraries there are several constraints applying to conventional journals which are lifted when the information is prepared and stored electronically. For example, the cost of storing an electronic paper is barely influenced by its length.

Cross-Linking and Cross-Referencing. Electronic publishing offers a number of useful linking and hyperlinking features. Jumping directly from a table of contents to the section of interest is the most basic tool. It is possible to move directly from a reference in a journal of one publisher to the paper cited, even though the latter may have a different publisher. You get the abstract for free. To read the full text you have to identify yourself as a registered subscriber or you are requested to pay per view. This new option helps you to get the information of interest immediately and to save a lot of time.

This initiative, known as *CrossRef*, is now supported by over 600 publishers. *CrossRef* itself is a nonprofit membership organization. It already provides links between millions of articles in thousands of journals. Further details can be found at:
http://www.crossref.org.

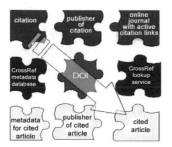

Fig. 6.5. Schematic illustration of cross-referencing

Unique Identification (DOI). Every digital object, be it a complete journal, an article, a book or a book chapter, can be assigned a Digital Object Identifier (DOI). This is a unique and persistent identification number.

The DOI itself consists of two parts: a prefix, which is uniquely assigned to each publisher by the DOI administration, e.g. 10.1007 for Springer-Verlag; and a suffix which is freely chosen by the publisher to identify the document or other electronic file concerned.

All DOIs and the corresponding URLs where the objects are located are registered in a central DOI directory that serves as a routing system. The underlying technology is based on the Handle System developed by the Corporation for National Research Initiatives.

To retrieve an object, the user sends the DOI to this directory and the request is forwarded to the associated URL. Either the object itself or a response page with further access information will be returned to the user. For example, by entering http://dx.doi.org/10.1007/s00214990m180 in a web browser, details will be delivered about a regular article that appeared in the journal 'Theoretical Chemistry Accounts'. The page displayed contains title, abstract and bibiliographic information. The article itself is available only to subscribers. For more details about DOIs and their use see http://www.doi.org.

Statistics. Electronic usage statistics can tell you how many times a particular publication is accessed.

Information for Third World Countries. Whereas, in the past, many poorer countries were virtually cut off from scientific information due to the high prices of journals and other publications, electronic publishing is helping to close this gap. A number of initiatives have been taken to provide free online information about scientific and medical research to poorer countries. An example is the '*Access to Research*' initiative set up by the World Health Organization with the support of six international scientific publishers.

In fact, it is not only readers who benefit from the above-listed features. They bring advantages for authors too. As author of an electronic publication you can be sure that your contribution will be found (provided you supply the correct keywords and/or subject codes); you can make use of supplementary material to convey your message; profit from a faster availability; and know that your work is embedded in a network of cross-linked information – the edifice that is scientific knowledge.

Coupled with the emergence of electronic publications, many journals have also developed more streamlined submission and refereeing processes, often taking place exclusively electronically. This too helps to simplify the job of authors. And editors, referees, and publishers also profit from this increase in efficiency.

Present Your Work in a Comprehensive Way. Whenever legally and technically possible you should post your papers on your own homepage and on your institute's homepage. You can even scan your older papers (pre-1996) and present them electronically. This small effort will provide benefit to the community and to you yourself.

6.4 The Role of the Publisher

All of us could bake our own bread if we chose to. But almost everyone prefers to go to the local bakers or supermarket to buy a loaf. There are good reasons for this: We prefer to leave the job to a specialist who produces a good product at an affordable price. And it is more convenient and often less expensive to buy mass-produced goods from a proven source.

This reasoning applies analogously to scientific publishing, and, in particular, to electronic publishing.

The numerous tasks and challenges involved in producing and maintaining a source of high quality up-to-date scientific information require a dedicated team of specialists with the necessary expertise and infrastructure. For electronic publications, many of the same tasks apply as for conventional paper publishing; but new ones are added. Let us give a summary of the major tasks involved in running an electronic publishing platform.

6.4.1 Tasks Involved in Electronic Publishing

▷ Provide and maintain the hard and software needed to offer a fast and flexible service.

▷ Organize publications in a framework of journals and book programs.

▷ Find and maintain teams of external reviewers and advisors to evaluate the quality of the information submitted for publication (peer reviewing).

▷ Acquire suitable material for publication.

▷ Observe international copyright rules.

▷ Ensure the formal quality of the publications (language editing, style of presentation, formatting, quality of illustrations).

▷ Make information available with a minimum of delay.

▷ Draw attention to the available information and products through appropriate announcements and other marketing activities.

▷ Archiving publications to guarantee long-term availability.

There is much debate in the scientific and publishing communities about the future of scientific publishing, with many doommongers predicting that publishers' days are numbered. However, a glance at the above list of tasks should make it clear that electronic publishing is hardly less demanding to organize than conventional publishing. All the tasks listed require specialist knowledge, organizational skills and – last but not least – have to be financed. Even a non-profit-making academic organization cannot offer a comprehensive scientific information service without outlay, significant organizational effort and substantial financial resources. In 2003 a study published by the American Association of Research Libraries showed that the average expenses for development and support per online journal of the scientific societies are 2.5 million US $ per year. In fact, some journals are going back to the earlier method of levying page-charges in order to finance their services.

6.4.2 Are Publishers Needed?

The publishers' standpoint, therefore, might be paraphrased by: Like bread-baking, surely the best way to maintain quality, uniformity and availability is to leave the job to the specialists. Whether the customer (reader/institute) or the originator (author/institute) should foot the bill, or whether the expenses should be shared is a matter of personal preference – one can find examples of all of these models. But one thing is clear: without organized and financially viable infrastructure, high-quality electronic publishing of scientific information is not possible.

6.5 Problems and Potential Pitfalls

Although electronic publishing is becoming well-established and accepted within the scientific community, it has yet to replace conventional publishing. This is because of a number of open questions that still have to be solved.

Long Term Availability. A risk attached to electronic-only journals concerns the availability of the information for future generations, i.e. the question of archiving and conversion. We all know how rapidly data formats and compression algorithms are replaced by new and better ones. Despite ongoing attempts to develop standards that will guarantee long term accessibility, there always remains the danger that the data of certain electronic publications will become irretrievable at some point in the future. Will an article published on the web still be readable in 10 years from now, or 100 years? Texts written on paper 2000 years ago can still be read today. Thus, even with the widespread use of electronic media, many libraries still like to have a print version of articles for archiving. The reader's priority is to have quick access to information via the PC but the author would also like to be sure that his work will remain available in the next century. The only way to reconcile these somewhat contradictory desires is the parallel publishing of print and on-line texts. The reluctance to accept electronic-only publications is clear from the relatively small numbers of papers being submitted to electronic-only journals.

Archiving. The topic of long-term availability also leads to the next question: Who should be responsible for archiving electronic information. The publisher or every library for itself?

Quality Control. The next problem area is quality control. In days gone by it was not easy for an individual to publish his or her ideas. The only way to disseminate information widely was to go to an international publisher and have them evaluate your work. If it was good, it would be accepted and would – hopefully – be widely read and gain recognition. These days anyone can publish on the internet. And, as anyone who uses Google will know, the internet is also a vast graveyard of digital junk. It is not possible to tell from a particular website what standing the organization behind it actually has. It can be a reputable scientific publisher or a group of lunatics with an axe to grind. Thus the matter of quality control is particularly important for online scientific publishing.

This is also another reason why it is better to concentrate scientific publishing activities among a few large organizations/companies willing to organize and maintain strict peer-reviewing of all material published.

Flood of Information. Another possible disadvantages should also be mentioned. There is the danger of a proliferation of material that exceeds the scientist's capacity to evaluate and assimilate it all. This must be avoided by continuing the peer-review selection process and attempting to remain selective about what material is published electronically. As for printed journals, it is likely that only a few major websites and electronic journals will maintain the quality and prestige needed to survive in the long term. This process of 'natural selection' is to be welcomed.

Evaluating the Impact of a Paper. The present science citation index gives a measure of the impact of a journal by determining the average number of times that its articles are cited.

How can this type of evaluation be extended to cover the readership of articles? For electronic publications at least, it is possible to record the number of times that each article is accessed, a statistic that could in future also be used to define a popularity ranking.

Pricing Policy. That electronic publication are not cheap to produce has already been demonstrated in the previous section. So far, however, there is no uniform scheme for financing electronic publishing.

There are several models in use, but the likelihood is that one will gradually emerge as the winner. Which it will be is hard to predict:

> ▷ access permitted only for paying subscribers and via pay-per-view (currently used for most electronic journals)

> ▷ free access to journals but significant page charges for publishing there (hence making it difficult for scientists from poorer countries to publish), see Sect. 6.7 on Open Access

ISIHighlyCited.com.
Powered by ISI Web of Knowledge   **Fig. 6.6.** ISI evaluates impact

▷ combination of moderate charges for authors and readers

▷ government funding (direct or via National Research Institutes) of electronic publishing. Argument against this: political interference could jeopardize scientific objectivity.

6.6 Future Prospects, Near and Medium-Term

The big question for the future is: Will electronic publishing take over as the only way of disseminating and storing scientific information?

It is difficult to make long-term predictions, but for the near future (next 10–20 years) we consider a scenario with the following features most likely:

Electronic publications will offer a useful extension of print media, but will not replace these entirely. Printed publications, especially books, will continue to be a sought-after commodity. It is certainly more convenient to read conventional textbooks on paper than on a computer screen.

What might change this reading habit in the medium term is the introduction of electronic paper – a medium satisfying the bed-bath-beach criterion which has always been cited as an argument in support of the conventional book (see Sect. 6.6.2).

Scientific journals and conference proceeding will be published predominantly in electronic form; the paper form will continue to be provided as an added extra for some time to come, since many libraries consider this necessary for archiving purposes.

Reference books too can look forward to a largely electronic future. They will be available either online or as a CD-ROM.

New forms of electronic publication will gain ground: e.g. virtual journals and '*jooks*' or '*bournals*'.

The use of electronic publications will become more convenient and efficient as CrossRef and other linking features become better established.

Copyright laws may be relaxed somewhat. It will be a challenge to find new copyright laws that continue to protect the intellectual property of the individual, also giving sufficient protection to publishers that they can continue to cover the costs of their service, but at the same time increase the efficiency of the publication process, especially with respect to online publications.

Scientific publishers will continue to provide a service to the scientific community. Their main tasks of quality control (peer reviewing), and organisation, formatting, and making available scientific information will not change significantly with the shift to electronic publications.

After an interim period in which many scientific societies, universities, institutes and companies endeavour to start their own electronic publishing channels, there is likely to be a period of consolidation, in which only the best of these survive. A similar process has already taken place in the field of internet retailing. At the end of the day it was mainly the online shops of the established high-street retailing companies that stayed the course. With the notable exception of Amazon, almost all of the new companies that were conceived, mainly during the new-economy boom of 1999–2000, as purely electronic retailers, have now disappeared again.

And now, having made some tentative predictions for the near and medium-term future, let us put these into perspective by looking at some earlier predictions about the future in which we are now living.

6.6.1 How Predictable Are Future Developments?

This question is most easily answered by considering some of the confident predictions made by experts in former times.

The consumption of paper was predicted to decrease dramatically with the advent of e-mail. In fact, it has increased by 100% within the last 20 years. E-mails are quick and easy to write and so many more get written than would similar letters. And these e-mails are then printed out with equal abandon for the filing cabinet.

The demise of paper checks was predicted for about 1970. In 1966 Americans wrote 20 million checks. Thirty years later, in 1996, this figure had increased to 64 million.

In 1893 the President of the American Patent Office predicted that there would be no new discoveries in physics after about 1900. He also proposed that the Patent Office should then be disbanded.

IBM President Thomas J. Watson, speaking in 1943, expressed his thoughts about the need for computers with the famous sentence: '*I think there is a world market for 5 computers*'.

Bill Gates, speaking in 1981 about the need for increasing computer power, said: '*640K ought to be enough for anybody.*'

Marshall McLunan, who was considered something of a guru in the publishing world, claimed in 1968 that we would see the end of the printed book in 1972.

Some people think that electronic publishing will one day mean that world-wide there is only one copy of each book or journal. But such a prediction could well join the ranks of the above fantastically wrong predictions. We, at least, shall make no such prediction.

6.6.2 Electronic Paper: A New Technology

Before concluding this chapter, we turn our attention to a subject that has been mentioned several times in passing, electronic paper, also known as digital paper or e-paper.

Fig. 6.7. Electronic paper: Bichromatic beads (Gyricon Media): impression and principle of realisation

E-paper is a very special technology that may one day result in a fusion of electronic and conventional publishing. In its physical appearance and properties it has much in common with normal paper: thin sheets bearing text and images. But it also takes over the role of a computer screen in that it can be fed with electronic data from a computer or another data carrier, even - via radio - from a remote source. The pages are only temporarily charged with information and - whenever the reader wishes - a new page of information can be loaded, in much the same way as a computer-user moves from one web page to another. Hence e-paper unifies many of the advantages of conventional paper and electronic displays, in particular portability and speed.

There are two main technological concepts for realising electronic paper:

(i) A cheaper concept, restricted to black and white displays, based on a thin plastic sheet containing millions of tiny beads, each half black and half white. Electric potentials are used to rotate the beads through 180 degrees, thus generating the display.

(ii) A more sophisticated technology allowing full-colour displays and making use of organic light emitting diodes.

Whereas the first concept begins with the idea of conventional paper and attempts to add on the facility of charging this with electronic data, the second starts at the other extreme: beginning from the concept of a conventional computer monitor, it attempts to turn this into a thin flexible sheet with paper-like properties. Both concepts are at an advanced stage and some products are already commercially available, in particular advertising signs based on the first variant.

In principle, electronic paper will be able to replace paper in virtually every application involving information display. In particular, it could be interesting for the delivery of books and journals in higher education and academia. One can imagine students in the future needing to own just one electronic book made of e-paper, and use this same set of pages to read all their course material.

Scientific journals and their readers may also profit from new delivery models. Rather than reading articles online or printing them out, students and researchers may choose to download them to e-paper. This enables them to read the material at their leisure: in the bath, on the beach or in bed – in exactly those places that conventional PC monitors can barely reach.

Whether or not e-paper will really take off is still difficult to say. Although it has obvious benefits in combining the advantages of traditional paper with those of modern electronic communication, there are several factors that may hinder its acceptance for general use. We end this section with a summary of the currently apparent advantages and disadvantages of e-paper:

Potential Benefits

High Quality Displays. The quality of e-paper is better than that of LCDs. Its resolution is higher and its contrast close to that of paper. The display can be read in direct sunlight and remains legible at quite wide angles.

Low Cost. The materials needed to make e-paper are inexpensive and the manufacturing processes is also cheaper than that for LCDs. The sheets can be erased and refilled with text more than a million times, so that a single book of e-paper can transmit as much information as anyone could expect to read in a lifetime.

Table 6.1 Present and predicted time-scale for development of e-paper

1999/2001	2003	2005	2010
first in-store displays installed	first displays for mobile devices	prototype books and newspapers	markets for books and newspapers become large-scale

Low Power Consumption. Once printed, the information on a sheet of e-paper is maintained with no power supply. One manufacturer of black and white displays E-Ink, claims that its e-paper needs about 1 % of the power used by a standard notebook computer screen. However, for full colour polymer displays the power consumption is higher.

Lightweight. E-paper needs no heavy bulky power supply. Once written on, it needs no power at all. This makes e-paper products much lighter than their monitor counterparts, comparable indeed with traditional paper products.

Flexibility. Manufacturers expect that e-paper will be truly paper-like and have the same flexibility in that it can be folded. This will certainly set it apart from other types of digital display.

Wireless Loading. Data can be received from a remote source via radio. This allows that rapid dissemination of news and other important changing information.

Making and Storing Notes. Present prototypes of electronic paper cannot themselves be written on. However, this facility is likely to be developed in the foreseeable future and will greatly increase the usefulness of the medium.

Disadvantages and Risks

There a still a number of problems and unknowns which pose stumbling blocks for the widescale introduction of e-paper.

Life Span. The present life span of an e-paper display is between 10,000 and 30,000 hours, compared to 100,000 for a PC monitor. This will have to improve if e-paper is to become competitive with other display technologies.

Batteries. Although e-paper requires little power, some form of power supply is needed for it to display new information. This may hinder its use in some applications of traditional paper.

Lack of Uptake by Manufacturers. So far, only Philips Components has agreed to invest in E-Ink technology. Other manufacturers of mobile displays, for example Nokia and IBM, are still very keen on improving LCD technology.

Consumer Acceptance. At the end of the day it will be the consumer who decides whether and when e-paper will become a household product. Ideally, the technology needs to find a 'killer application' in order to show off all its benefits and make it known to a wide cross section of the community. It is not clear who will be most attracted to electronic paper: it could be those who dislike computer technology and have thus far steered clear of electronic information. But it could equally well be the gadget lovers and possibly both groups.

6.7 Open Access Publications

Open Access (OA) is a new business model – in fact a whole new philosophy – in publishing and in the dissemination of information. It is currently a topic of heated debate, especially in scientific publishing and academic circles. OA literature is digital, online,

Fig. 6.8. Open access journal

free of charge to the reader, and may also be free of most copyright and licensing restrictions.

OA initiatives are especially prevalent in science, where it is clear that free access to the latest information will help to promote progress in research. Hence, many universities, institutes, libraries and even the British House of Commons are actively encouraging the expansion of OA scientific publishing.

The main difference from traditional publishing is that the costs for preparing and disseminating OA literature are not borne by readers or their libraries, and hence to not pose access barriers. So if the readers are not paying, who does finance OA publications and what are the costs involved? When one considers peer-reviewed, formatted, and (where necessary language-edited) articles of the type published in typical scientific journals, it turns out that the true cost – including IT costs – of making each article available lies somewhere in the range \$ 2,000 to \$ 7,000. The American Physical Society, for example, showed that distributing all the costs incurred would lead to a \$ 4,000 tag for each article published in Physical Review B. Some of the first OA initiatives were, and are still being, financed by institutions, universities, public libraries and other donors. But the model now being introduced by many large publishers (both commercial and non-profit) is one in which the authors of the articles are themselves asked to bear some or all of the costs [Déjà vu? All older scientists will remember page charges; really this is just an extreme form in which the page charges are so high that the subscription rate can be reduced to zero.]

At the time of writing, there are only a handful of scientific journals that are 'pure open access'. Examples are the *New Journal of Physics*, jointly published by the Institute of Physics and the German Physical Society; and the Oxford University Press Journal *Nucleic Acids Research*. In both cases, authors pay to have their articles published. A journal in which publications are free, both to readers and to authors is the *Dermatology Online Journal*, which is sponsored by the University of California and the California Digital Library. One of the earliest OA journals, it was first published in 1995.

At present it is not clear whether OA will one day completely take over from traditional subscriber-based scientific publishing. To keep all their options open, some publishers, including Springer, are offering elements of both models within the same journal. Springer's system, known as '*Open Choice*', allows authors, once their paper is accepted for publication, to choose whether, in addition to the normal publication in the subscriber-based print and online journal, they would like to pay for their paper to become open access as well.

Despite its obvious advantages for the scientific community, OA also has its pitfalls. One of these is the temptation to be less strict about scientific quality and to relax standards of refereeing, since more papers mean higher income. But this can be avoided by having a truly independent refereeing process; furthermore, market forces will – as they do now – lead to the good papers being concentrated in the good journals, a fact that will stimulate publishers to maintain standards. Another drawback of the '*author pays*' model is the fact that some researchers, especially in third-world countries, may not have the funds available to publish in good journals. But on the other hand, third-world libraries are often lacking the funds to subscribe to high-priced conventional journals. So this a 'swings and roundabout' situation.

Only time will tell which model will assert itself in the long term. Perhaps both can co-exist; perhaps another completely different scheme will be devised. Certainly these are interesting times for the publishers, authors and readers of scientific journals.

7 Patents

Well informed people know that it is impossible to transmit the voice over wires, and that, were it possible to do so, the thing would be of no practical value.

Editorial in the Boston Post, 1865

7.1 Introduction

The editorial quoted above was of course referring to Alexander Graham Bell's invention of the telephone. For this invention, US patent No. 174,465 was issued in 1875, now regarded as the most valuable patent of all time. Although your invention may not change the world quite so dramatically as the technology of the telephone has done, the possibility of obtaining a patent is something that – as a scientist or engineer – you should always keep in the back of your mind.

Successfully applying for a patent is a complex procedure fraught with many pitfalls. It is also relatively expensive. But if your idea, or rather the practical invention based on it, is truly new and has the potential to earn money, then you should seriously consider protecting it with a patent before making it publicly known. Your first step should be to obtain confidential professional advice from a patent attorney or from the patent department at your company or institute. This professional advice should tell you: (1) Whether your invention is patentable; (2) whether it is worth patenting and if so where; and (3) what the whole procedure is likely to cost. If your organization will apply for the patent on your behalf, you should also familiarize yourself with your rights to a share in the income generated by the patent, e.g. through inventor reward schemes.

Fig. 7.1. US patent

In the remainder of this chapter we will give some basic guidelines about patent protection, how it works, how a patent is obtained, what costs are involved and, finally, some tips for seeking further advice and information.

The idea of patents for inventions goes back many centuries and its origin is rather obscure. No one country can claim to have been the first to introduce a patent system. However Britain is the country that has the longest continuous patent tradition. Its origins can be traced to the 15th century, when the Crown started making specific grants of privilege to manufacturers and traders. Such grants were signified by '*Letters Patent*', which were open letters marked with the King's Great Seal. The earliest known English patent for invention was granted by Henry VI to Flemish-born John of Utynam in 1449. It gave him a monopoly for a method of making stained glass, required at the time for the windows of Eton College, and not previously known in England.

7.2 What Is a Patent?

A modern patent is a legal document granted by a national government to an inventor. It gives you, the inventor, the right – for a limited period of time – to stop others from making, using or selling the invention without your permission. If you are awarded a patent on a particular invention, that invention becomes your property. In common with other forms of property or business assets, a patented invention may be bought, sold, rented or hired. Patents invariably cover only certain territories, usually individual countries. If you hold a patent for a particular territory, it gives you the right to stop others from manufacturing or selling the invention within that territory and also the right to stop anyone from importing it. If someone infringes these patent rights, it is up to you, the patent-holder, to take action, e.g. by seeking an injuction and damages. However, it is often better to try and reach an agreement with the infringer before resorting to legal action.

Patents are usually valid for a maximum period of 20 years from the date of filing the application.

7.3 What Can and Can't Be Patented?

Not all types of invention can be patented. And even for those that can, certain conditions have to be fulfilled (see Sect. 7.4).

Generally speaking, patents are intended to cover products, apparatus, or processes that involve new functional or technical aspects and are capable of industrial application. Thus they relate to how things work, what they do, how they do it, what they are made of or how they are made.

Thus it is possible to patent, for example:

1) A machine (e.g. a mechanism with moving parts)

2) A manufacturable article (e.g. a hand tool)

3) A composition of matter (e.g. a plastic or alloy)

4) A process (e.g. a method of semiconductor fabrication or a chemical synthesis)

5) A new use or improvement of an existing invention (e.g. a new use of a known drug).

The following, however, are not patentable:

1) A discovery (i.e. something that was always there, although formerly unknown)

2) A scientific theory or mathematical method

3) Other ideas (e.g. ideas for games, although the technical components of a game may be patentable)

4) An aesthetic creation such as a work of art or literature

5) A computer program (here there are rare exceptions which depend on what the program does)

6) Anything with a moral or ethical impediment (e.g. a land-mine or letter-bomb).

The patenting of animal and plant varieties, genes and other biological material is currently a legal grey area. If you are interested in patenting a biological invention involving living matter it is wise to seek expert advice as early as possible.

7.4 Conditions for Patentability

Even though your invention may fall into one of the above categories of patentable inventions, it will only be granted a patent if it fulfils some further conditions.

A patentable invention must:

1) Be New

For most patent authorities (US excepted, see Sect. 7.7) an invention is considered new only if it has never been made public in any way, anywhere in the world, before the date on which the patent application is filed.

2) Involve an Inventive Step

An inventive step means that your invention, when compared with what is already known, is something that would not be obvious to a person with a good knowledge and understanding of the subject.

3) Be Capable of Industrial Applications

This means that the invention must have some form of practical application as an apparatus, device, product or process. 'Industrial' is to be understood here in its broadest sense as anything distinct from purely intellectual or aesthetic activity. It need not imply the use of machines or the manufacture of an article. It also includes agriculture.

4) Be Consistent with Known Physical Laws

Patent offices throughout the world continue to receive applications for patents on perpetual motion machines, teleportation devices (beam me up Scottie), and time machines. These all fail the patent examination – in fact, only on a technicality. A patent can only be granted on inventions that can also be used by other suitably trained persons!

7.5 Who Should Apply? Patent Ownership

If you have made an invention, then according to patent law, you are the first person having the right to apply for a patent. In practice, however, it is often a company or institute that applies for the patent. But the name(s) of the inventor(s) must always be given on the patent and these are always natural persons and not a company.

In theory, as an employee who makes an invention you are considered to be the first owner of any patent rights, unless you were expressly hired to invent, or have explicitly assigned your invention to your employer. But in practice – and as confirmed by numerous court cases – an implied assignment of an invention is often deemed to exist when the invention is made during the course of employment. Factors that would support an employers right to obtain a patent on an invention include:

1) Use of company funds to do R&D on the invention

2) Development carried out in company time

3) Inventor hired to perform research, engineering or experimental work

4) Inventor is a director or officer of the company.

In contrast, factors that would support an inventors right to obtain a patent include:

1) Development begun prior to, or continued after period of employment

2) Development pursued with inventors own funds and/or equipment

3) Inventor raises funds for development and/or markets invention

4) Invention is not related to company business or to employee's specified job.

Fortunately, most employee-inventors do not go away empty handed. Many companies have inventor-reward schemes which give employees a share in the earnings from patented inventions, licences etc. An inventor who does not feel that he has been fairly compensated, can also resort to legal action to claim his share. In a recent high-profile case Shuji Nakamura, inventor of the blue light emitting diode, was awarded 20 billion Yen ($ 185 million) by a Japanese court. His company, Nichia, had only paid him a bonus of 20 thousand Yen ($ 185) in recognition of his technological breakthrough. The ruling issued by the Tokyo District Court stated that Nakamura's invention *made possible the manufacture of the blue LED*' and estimated that Nichia would earn 120.8 billion Yen in profits from the exclusive rights it holds on the device up to October 2010, when the patent expires.

If you are unsure about who should apply for a patent on your invention – or are not satisfied with the suggestions of your in-house patent department – you should seek the advice of an independent patent attorney.

7.6 Before You Apply

If you believe that you have an idea, or have made a discovery that is, or can lead to a patentable invention, you should seriously consider protecting this piece of intellectual property by applying for a patent. In this way you – and your company/institute – can hope to benefit financially by licencing the invention or directly selling products that exploit it. But be sure that there is really a market for your invention/product and that someone is willing to advertise and sell the product once you have applied for the patent. Of course there is nothing to stop you from developing and selling a new product or technology without a patent. But if you do this, and if the product is a success, you must expect imitations to appear in a short time.

Before spending a lot of time and money to obtain a patent, we recommend that you (a) read one or the other of the two books given in the footnote below,* and (b) seek confidential professional advice (a nondisclosure agreement should be signed before you share details of your invention with any third parties.)

Next, or better still in parallel, you should convince yourself that the idea is really new (literature search!) and that the invention actually falls into one of the categories of patentable inventions. A detailed search is something that can be left to a patent attorney (and

---

* Richard Levy: *The Complete Idiot's Guide to Cashing in on Your Inventions* (Alpha Communications, 2001)

  James E. White: *Will It Sell? How to Determine If Your Invention Is Profitably Marketable* (Before Wasting Money on a Patent) (James E. White and Associates, 2000)

later the Patent Office will also perform a search), but the initial search you can do yourself.

The websites* of the main patent offices offer help and links to databases for search purposes.

Until a patent application has been filed, it is wise not to discuss your idea with other people and it certainly should not be made public by disclosing it in a talk, a paper, or on the internet. Throughout Europe and in many other countries, prior publication of the idea will prevent it from being patented. In the US, the situation is a little more lenient, allowing inventors 1 year to apply for a patent after publishing their idea.

Although, as a scientist, you will probably be skilled at describing the scientific and technical aspects of your invention, applying for a patent can be a daunting procedure. It requires, in addition, extensive legal knowledge and expertise in formulating legal documents, in particular in relation to the claims made in the patent. There is also considerable paperwork to be dealt with and deadlines to be kept for payments. Unless you are familiar with these matters, or have plenty of time to study them, it is strongly advisable to seek professional assistance when applying for a patent. Many institutes and most large companies have a department dedicated to technology transfer and/or patenting. If such a facility is not available to you, the best alternative is to seek the services of a registered patent attorney or patent agent. But don't choose one at random from the internet – try to get a recommendation from a reputable source.

In deciding where to apply for patent protection, you and your legal advisor need to consider the potential market for the product, device or process that you wish to patent. Another criterion may be the funds available – the larger the geographical region in which you seek patents, the more it will cost you (see Sect. 7.10). Many inventors satisfy themselves with a single patent in their own coun-

---

*   websites of the major patent offices:
www.epo.org (European Patent Office)
www.uspto.gov (United States Patent and Trademark Office)
www.jpo.go.jp (Japanese Patent Office)
www.patent.gov.uk (United Kingdom Patent Office).

try, relying on their technological lead to earn money from their idea elsewhere. The world's three biggest patent offices are the European Patent Office (EPO), the United States Patent and Trademark Office (USPTO), and the Japanese Patent Office. Each issues patents for its own territory. Via the EPO you can apply for patents in as many of the 27 member countries as you wish.

7.7 Patent Application Procedure – Example of European Patent Office

Here we will give a summary of the application and examination procedure employed at the EPO. The procedure and rules at other patent offices differ in some details, if anything being slightly simpler. (In Sect. 7.9 we describe the main differences between European and US patents).

A European Patent Application must be filed at one of the EPO offices in Munich, The Hague or Berlin. The application itself can be submitted in English, French or German.

The application usually consists of five parts:

1) A Request for Grant. This is a formal request for a patent to be granted and must be submitted on an official form available from the EPO and affiliated patent offices.

2) Description. This must provide a clear description of your invention – in relatively nontechnical language – which would enable any person skilled in the relevant area to make and use the invention.

3) Claims. The claims define the aspects of the invention, and the technical features, for which patent protection is sought. They must be clear and concise and be supported by the description.

4) Drawings. These are optional, but are frequently the best and most convincing way of illustrating the advantages and novel features of your invention.

5) Abstract. The abstract contains purely technical information and cannot be used to assess the patentability of the invention.

It takes considerable skill and experience to draw up a patent application in a form that fulfils the various legal and formal requirements and also gives you maximum protection for your idea. Thus this is usually the job of a patent attorney. Often the inventor will write up a less formal '*inventor's report*' and submit this to the company's, or an independent, patent attorney to draft the official application.

Once the patent application has been filed and the date of filing confirmed, it is safe to publish the idea. At this stage one can begin to mark products that use the invention as '*patent pending*'. This does no more than inform potential competitors that a patent has been applied for. They remain free to copy the idea until such time as the patent is granted. However, the prospect of then having to pay licence fees, or worse still withdraw their product, may deter them from the outset from investing in a competitive product. This is the real reason why companies like to mark their products as '*patent pending*'.

7.8 The Patent Examination Process (EPO)

Having filed your patent application, you will then have plenty of time to turn your attention to other matters. Although you and/or your patent attorney may be called upon to communicate with the patent office on occasion, it is likely that you will hear little or nothing for the next 18 months. In fact, the typical time between filing an application and the granting of a patent is about 3 years (paradoxically often because the inventor wants to delay the issue of a patent for strategic reasons). What happens during this time?

1) The first step after filing is a '*Filing and formalities examination*' to establish that you have provided all the necessary information, documentation and fees.

2) In parallel with the formalities examination a search report is drawn up, listing the documents available to the EPO which may be used to assess the novelty and inventive step of your idea. At

Fig. 7.2. The appeal process

this stage no opinion is given as to whether your invention is patentable. The search report will be sent to you together with any cited documents.

3) The patent application is published – normally together with the search report – 18 months after the date of filing. You then have a period of six months in which to decide whether to pursue the application. Of course, if it turns out that one of the documents found by the EPO describes an exact replication of your invention, you would be wise to cut your losses at this stage. If, however, the documents found in the search give no reason to doubt the novelty and inventive step involved in your idea, you should request that the EPO proceeds to undertake a '*substantive evaluation*'.

4) The substantive evaluation. In this process the EPO examines, in the light of the search report, whether the application and the invention to which it pertains meet the requirements of the European Patent Convention and, in particular, whether the invention is patentable. The final decision – which often follows extensive discussion and modifications to the claims – is made by a panel of three patent examiners. If the applicant is still not satisfied, an appeal process is possible.

5) The patent is approved. What is actually granted by the EPO is not one European patent but a bundle of national patents. These have to be translated, where appropriate, into the languages of

the countries concerned and then filed with the national patent offices. Thus while the EPO application procedure is lengthy and complex, it has the advantage of gaining you patents in as many as 27 European countries. You are free to choose the countries for which you seek patent protection. If three or more countries are concerned, then an application via the European Patent Office is worthwhile.

7.9 Differences Between US and European Patents

Although many aspects of patent law have been harmonized internationally through treaties such as the Patent Cooperation Treaty, there are still a few important differences between the two major systems.

First to File Versus First to Invent. If two people apply for a patent on the same invention, the EPO will grant the patent – assuming the invention proves to be patentable – to the first person to have filed an application. The order in which the applicants made the invention is irrelevant. In the US, however, the criterion for awarding the patent is the order in which the inventions were made. This often means that laboratory books have to be examined, or the dates on which prototypes were made must be established. Thus if you believe you may one day want to obtain a US patent, should be very careful about recording and dating all steps in your research and development work.

Grace Period. The European Patent Convention (EPC) demands that an invention should not have been publicly available prior to the filing of a patent application (Article 54 EPC). '*Publicly available*' includes selling the invention, giving a lecture about it, publishing it in a journal and showing it to an investor without a nondisclosure agreement. In the US, the law is more lenient: Inventors have a period of one year after publishing an invention in which to apply for a patent. But any Americans wishing to apply for European (or Japanese) patents should take care to keep their inventions secret until such time as the patent application is filed.

Best Mode Requirement. US patent law requires you to include on the patent application the best way to put the invention into practice. This prevents inventors from gaining a patent whilst still keeping some essential or advantageous aspect a secret. European law, however, has no such requirement, stating only that the application must specify at least one way of practicing the invention. It need not be the best way, and need not even be a good way.

Territories Covered. A US patent entitles the holder to prevent anyone from making, using or selling the patented invention in the whole territory of the USA. A European patent conveys the same rights to its owner, but only covers a selection of countries as specified by the applicant. These can be any or all of the 27 countries that are party to the European Patent Convention.

There are other minor differences in the exact definition of patentable inventions, in the procedure for opposing patents, and in the form in which patent claims are specified. But these would take us into an unreasonable amount of detail and are – we trust – matters that a compentent patent attorney can deal with on your behalf.

Japanese patents, incidentally, are awarded according to much the same rules as in Europe. First to apply takes precedence over first to invent; and a criterion of strict novelty applies – i.e. no prior publication is possible.

7.10 Costs of Patents

Asking what it costs to patent an invention is like asking '*How long is a piece of string?*' The answer is: It depends. It depends on the complexity of the invention, the patent authority to which you apply (i.e. geographical region covered), the strength of the claims you make, and – to the greatest extent – on the amount of professional help that you require.

The basic fees levied by the European and US patent offices are listed in Tables 7.1 and 7.2. But these are only the bare bones. Any delays caused by the applicant, any minor modifications to the submitted application, anything more than a sneeze and you will be expected to pay additional fees. In fact the official tables of fees have over 60 (Europe) and over 120 (US) different entries covering all eventualities, not forgetting the use of photocopiers. The largest group of fees are the penalties for late payment of fees due!

The very least that you would have to pay to be granted a US patent is thus \$ 1050 (filing fee plus issue fee), and for a European patent 3189 Euro (filing, search, examination, and grant). But it is a rare event that a patent application goes through for these minimum amounts. However, one can draw the generally valid conclusion that a US patent is significantly cheaper than the correspon-

Table 7.1. Basic fees for a US Patent in US \$

Payment	Large entity*	Small entity*
Provisional Application	160	80
Basic filing fee	770	385
Additional claims	86	43
Issue fee	1330	665
Maintenance fees		
At 3.5 years	910	455
At 7.5 years	2090	1045
At 11.5 years	3220	1610
Appeal	min 330	min 165

* A small entity is any legal entity that employs less than 500 people and a large entity is any legal entity that employs 500 or more people, including part-time workers.

Table 7.2. Basic fees for a European Patent in Euro

Payment	Amount (Euro)
Filing fee	125 (online 90)
Search fee	690 (European)
	1550 (International)
Fee per country covered	75
Examination	1430
Fee for grant	715
Opposition fee	610
Appeal fee	1020
Protest fee	1020
Maintenance of patent (annual fees up to 20th year)	different for each country covered

ding European patent, which typically costs about three to four times as much. The EPO itself claims that the fees it receives for an *'average'* European application (covering 8 designated states) and where the patent is granted after four years, amount to 4300 Euro. This includes the filing, search, designation, examination, grant and printing fees.

In addition, you will also have to consider the cost of professional help in drafting and submitting the application. Here too rates vary significantly, largely in proportion to the complexity of the patent.

Table 7.3 gives a very rough guide to the total amounts you can expect to invest – including attorney's costs – in order to gain a US patent on your invention.

Table 7.3. Rough amounts required to gain a US patent

Simple invention	$ 5,000–7,500
Invention of minimal complexity	$ 7,500–12,500
Moderately complex invention	$ 12,500–20,000
Relatively complex invetions	$ 20,000–30,000
Complex invention	$ 30,000 upwards

7.11 Getting Assistance

If you are working for a company or institute and are content with the professional advice offered by your patenting or technology-transfer staff, then you can leave everything up to them. However, if you are working alone or are not satisfied with the deal offered by your company, you may like to seek independent legal advice. As a first step it may be good to read the book Patent it Yourself (Nolo Press, 2002) by David Pressman, himself a patent attorney. But then do not patent it yourself. The chances of successfully and quickly gaining a strong patent alone are slim and there are plenty of documented cases of inventors who have lost out on rights to their inventions due to inappropriately formulated claims or other technicalities that an attorney would have been able to avoid.

Although more expensive, we recommend you to use the services of a qualified patent attorney rather than turning to one of the numerous enterprises bearing names such as 'Idea Promotion Agency' or 'Invention Submission Company'. Many – although not all – of these are less than scrupulous and you don't want to join the long list of inventors who have had to learn this the hard way. Told that their invention is going to change the world and make a fortune, many inventors, blinded by their attachment to their idea, have been taken for an expensive ride.

Assistance is also offered by what are known as 'patent agents'. Although they have usually passed the same exams as patent attorneys, they will not be able to represent you in court in the case of litigation or later infringement of your patent rights. But depending on the funds available, you may prefer to use an agent rather than an attorney.

If possible, you should seek personal recommendations for a good patent attorney or agent. If none is available, here are some simple criteria for recognizing a good patent attorney:

1) They should be able to understand the technicalities of your invention and express these lucidly in words for a nonspecialist.

2) They will make sure that your patent has broad claims, but not so broad as to be meaningless. Obvious or frivoulous claims will not be included.

3) They will prepare the patent application with a view to anticipating and winning any future infringement case in court.

4) Although most patent applications are rejected once if not twice before acceptance, a good attorney will quickly know what changes are needed in order to get it accepted.

Having located a few attorneys/agents it is a good idea to call them up and interview them briefly. Explain that you are seeking assistance to file a patent and ask them if they are willing to give an initial consultation for free. Some will, some won't. Find out on the phone, or at an initial consultation about their experience, their special areas of competence, and their fees and billing terms (how high is their deposit, what is their hourly rate, what do they charge for the initial filing, what are the likely total costs up to issue of the patent?). Together this should give you enough information to choose a good representative.

7.12 A Little Light Relief

People may tell you that your idea is crazy. In itself, this is not a reason for giving up on it – remember what they said about the telephone. In a similar vein, Lord Kelvin, president of The Royal Society said, in 1895, 'Heavier-than-air flying machines are impossible'. It was just 8 years before the Wright brothers proved him wrong. And in May 1906 they were granted US Patent No. 821,393, for their 1903 flying machine.

But just because an invention can be patented, does not mean that it is not crazy! Figures 7.3–7.8 show a few examples of some of the wackier things that people have tried to patent, or indeed succeeded in patenting.

Fig. 7.3. US Patent No. US 4233942. A device for protecting the ears of a long-haired dog from becoming soiled by food while it is eating. A tube contains and protects each of the dog's ears. The tubes are held away from the dog's mouth and food as it eats

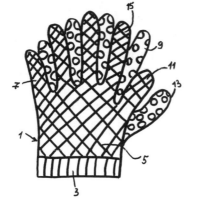

Fig. 7.4. UK Patent Application No. GB 2221607. A glove for courting couples who wish to maintain palm-to-palm contact while holding hands. It has a common palm section, but two separate sets of fingers

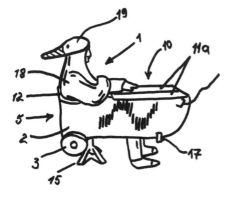

Fig. 7.5. Duck Decoy. US Patent No. US5687643 (issued 1997). Ducks can be very smart so, if you want to shoot a duck, you need to think and look like a duck. Quack, quack

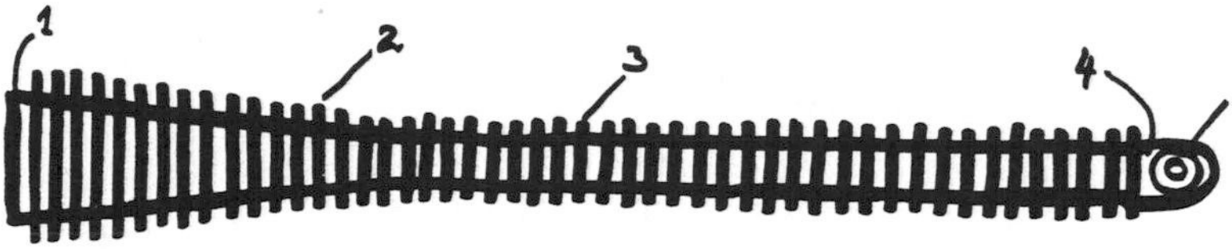

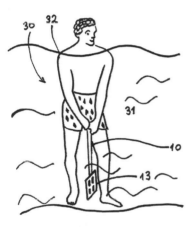

Fig. 7.6. UK Patent Application No. GB2272154. A ladder to enable spiders to climb out of a bath. It comprises a thin flexible latex rubber strip which follows the inner contours of the bath. A suction pad (5) is attached to the top edge of the bath

Fig. 7.7. US Patent No. US6325727. A device for training your golf swing underwater. The hydrodynamically adjustable paddle (13) can be altered manually, providing a variable resistance as the device is swung through the water

Fig. 7.8. US Patent No. 221,855, issued in 1879. Back in the olden days, when modern fire escapes were yet to be built, it was every man for himself. According to the inventor of this equipment '... a person may safely jump out of the window of a burning building from any height, and land, without injury and without the least damage, on the ground'. Perhaps he should have asked a physicist to check this ...